INTERNATIONAL CENTRE FOR MECHANICAL SCIENCES

COURSES AND LECTURES - No. 305

SELECTED PAPERS ON THE TEACHING OF MATHEMATICS AS A SERVICE SUBJECT

EDITED BY

R.R. CLEMENTS
UNIVERSITY OF BRISTOL

P. LAUGINIE
UNIVERSITE' DE PARIS-SUD

E. DE TURCKHEIM
INRA

SPRINGER - VERLAG   WIEN - NEW YORK

Le spese di stampa di questo volume sono in parte coperte da contributi
del Consiglio Nazionale delle Ricerche.

This volume contains 7 illustrations.

ISBN 3-211-82056-6 Springer Verlag Wien-New York
ISBN 0-387-82056-6 Springer Verlag New York-Wien

PREFACE

This collection of papers results from the ICMI Study, "Mathematics as a Service Subject". This was the third study which the International Commission on Mathematical Instruction (ICMI) has mounted. Previous studies have been on "The Impact of Computers and Informatics on Mathematics and its Teaching" and "School Mathematics in the 1990's". The present study is the result of cooperation between the Committee on the Teaching of Science of the International Council of Scientific Unions (ICSU-CTS) and ICMI. It culminated in a symposium held in Udine (Italy), from 6 to 10 April, 1987, at the International Centre for Mechanical Sciences (Centre International des Sciences Mécaniques, or CISM).

The study began by a careful investigation about the way mathematics is taught to students in another major subject in a few typical universities: Eindhoven Technical University in the Netherlands, Jadavpur University of Calcutta, India, Eötvös Lorand University and several other institutions in Budapest, Hungary, Florida Agricultural and Mechanical University in the USA, University College, Cardiff, UK, University of Southampton, UK, Université de Paris-Sud à Orsay, France. The past and current presidents of ICSU-CTS (the physicist Charles Taylor and the biologist Peter Kelly) took part in the program committe, which included also the president and the secretary of ICMI, the mathematicians Tibor Nemetz and Fred Simons, the statistician Elisabeth de Turckheim, and the physicist Pierre Lauginie. The Program committee issued a discussion document, which was circulated to all national representatives of ICMI, and to various institutions. It was published in the journal L'Enseignement Mathématique, tome 31 (1986), pp. 159-172, and it also appeared in French, Italian and Spanish versions. Abstracts or quotations appeared in other scientific or vocational journals. It was discussed among members of several scientific institutions (including the Académie des Sciences de Paris) and among professionals, for example the Fondation Bernard Grégory. Contributions to the discussion were received from many countries and these formed the basis of discussions at Udine.

The meeting in Udine was attended by 37 participants from 18 different countries, on invitations issued by the program committe. The generous hospitality of CISM – located in a beautiful historical mansion – and the working atmosphere, made this symposium both pleasant and profitable. A general summing-up of the meeting, four main reports – by J.M. Bony (France), H. Murakami (Japan), H.O. Pollak (USA) and F.H. Simons (The Netherlands) – and four other selected papers can be found in Mathematics as a Service Subject *(Cambridge University Press), the third volume in the ICMI Study Series.*

This volume supplements and extends that in the Study Series. The first papers, many especially written after the Udine meeting, describe the situation relating to service courses in a number of different countries – developed and developing. There then follows a selection of papers intended to illustrate aspects of the problem in more detail, and to describe some responses which have been made. Finally we reprint the statement issued at the end of the Udine meeting.

In this the importance of service teaching both for the livelihood of mathematicians and mathematics is stressed, and a call is made for more resources to be devoted to it and for greater recognition to be paid to those mathematics teachers who give it their especial attention. It is a vital sector of mathematics education and the growth of 'continuing education' will make it even more important.

Financial help to mount the meeting was received from UNESCO, ICSU, IMU, (International Mathematical Union), CISM, the Royal Society, the Ministére de l'Education Nationale of France, IBM-Europe, IBM-France, and many universities or institutions which contributed to the expenses of participants. We sincerely thank all of them, and we hope that the success of this study will prove, once again, that international actions of this type meet a real need and have an important effect.

Finally, we should like to express ICMI's particular indebtedness to Elisabeth de Turckheim, Dick Clements and Pierre Lauginie who have acted as editors of this volume.

J.P. Kahane
(President, ICMI)
A.G. Howson
(Secretary, ICMI)

CONTENTS

Page

MATHEMATICS AS A SERVICE SUBJECT

A.G. Howson, J-P. Kahane, P.J. Kelly
P. Lauginie, T. Nemetz, F.H. Simons
C.A. Taylor and E. de Turckheim

ABSTRACT

This paper is the discussion document which was issued by
the Planning Committee as a starting point for the ICMI
study on 'Mathematics as a Service Subject'.

† This paper is reproduced from L'Enseignement Mathématique,
  32 (1986), 159-172 by kind permission of the editors.

THE REASONS FOR THIS STUDY

Since it was established in 1908, ICMI has always, and rightly, paid considerable attention to the problems which arise when mathematics is taught to students who are primarily engaged in studying other subjects. As early as 1911 a meeting was held on the theme "What mathematics should be taught to those students studying the physical and natural sciences?" (see L'Enseignement Mathématique, 13 (1911), 481-496). At the International Congress of Mathematicians held the following year, in 1912, there was a discussion on mathematics for engineers, and who should best teach it. Without doubt the questions of 'service mathematics' should always command ICMI's attention.

Nowadays the teaching of mathematics is much more widespread and varied than it was in the 1910s; indeed, than most people, including many engaged in its teaching, imagine.

All the scientific disciplines and many of today's businesses and professions demand a certain mathematical knowledge and understanding. At the university level an important part of mathematics teaching is that intended for students of other disciplines. It is this 'service' teaching - interesting, important, valuable, but poorly understood and analysed - which is the subject of our study.

The problems are many. They relate, for example, to the nature of the discipline employing the mathematics, to the 'language' of the user, and the manner in which the mathematics is used; they have implications for the education of senior high school students. Of necessity, the responses to the problems will differ in different countries and institutions, for specialists in the 'major' discipline and their mathematical colleagues will exercise different degrees of control over the formulation of syllabuses and the teaching of courses. Yet everywhere, evolution in the mathematics taught and in methods of teaching is rapid. Who teaches what, and how, and why? What developments can be foreseen?

ICMI and ICSU-CTS (the International Council of Scientific Unions' Committee on the Teaching of Science) decided to mount a joint study in the hope that it would produce a confrontation of all points of view from which a deeper understanding and improved practice might emerge. We are asking users (specialists in a variety of disciplines, students, employers) to reflect on their real needs, and to attempt to identify their objectives in teaching and learning mathematics. We are asking those who teach service mathematics, whether or not they are mathematicians†, to consider how their teaching should be adjusted to cope with new developments and techniques both in mathematics and in their student's major subjects.

It is hoped that in addition to improving the teaching of mathematics as a service subject the study will help reinforce cooperation between mathematicians and non-mathematicians. Finally, we hope that our considerations will also prove of value to those involved in teaching mathematics at a pre-university level.

THE ORGANISATION OF THE STUDY

In broad outline the study is being organised in stages similarly to those employed in the study on 'The influence of computers and informatics on mathematics and its teaching'‡. The first phase of the study took place in 1985. An informal questionnaire was prepared which sought information on such matters as: the present situation (In which disciplines is mathematics explicitly taught? How are the syllabuses determined? Who teaches the courses? Is mathematics used as a means of selection (and elimination)? Are there differences in the way 'service' mathematics is taught and assessed? Are there significant differences in the way mathematics is taught to different disciplines?); perspectives (In an ideal world who should teach service courses? How? When? What? What are desirable developments? What is the role of mathematicians vis-a-vis that of other specialists? How do students derive motivation? How can one introduce examples and applications? How serious an obstacle is language/jargon/symbolism? What are the possibilities for the integration of service teaching (over several disciplines)?); experiences and lessons drawn from

---

† For the purpose of this paper we shall use the term 'mathematician' to describe someone attached to a Department of Mathematics or who would consider his/her main academic field of interest to be mathematics.

‡ The Proceedings of the symposium on this theme held in Strasbourg in March 1985 have now been published as the first volume in the ICMI Study Series by the Cambridge University Press (ISBN 0 521 32402 5 hard cover, 0 521 31189 6 paperback). A volume of fifty 'supporting papers' which were submitted to the study can be obtained (price FF100) from Dr F Pluvinage, IREM, 10, rue du General Zimmer, 67084, Strasbourg, France.

particular innovations in service teaching. Detailed
responses were received from a dozen disparate institutions
in England, France, Hungary, India, the Netherlands, the USA
and Wales. These reports were submitted by individuals†
rather than by institutions. As a result they may well
reflect personal biases or even contain some inaccuracies.
Nevertheless, their great value lies in the different views
they contain and the variety of practices they describe. It
is on the basis of these contributions that the present
document has been written. ICMI and ICSU-CTS are most
grateful to all these contributors.

The second phase which will occupy 1986 begins with the
publication of this discussion document. The Planning
Committee - comprising the authors of this text - wish to
receive papers written on one or more of the themes
described below. Such papers, typewritten and not exceeding
16 pages in length, should be submitted to A G Howson and J-
P Kahane‡ before the end of 1986 (and preferably before 30
October). Those contributions accepted by the Planning
Committee will then be collected together and distributed
prior to an international seminar to be held at the
International Centre for Mechanical Sciences, Udine, Italy
from 6-10 April, 1987.

The number of places available at this seminar will be
limited and invitations to attend will in general be issued
on the basis of these preliminary written contributions.
The meeting itself will be given over to the presentation of
a small number of invited reports and the discussion of
salient points arising from the 'supporting papers'. The
publication of the Proceedings of the Udine meeting,
together with amended versions of the contributed papers,
will mark the end of the international stage of the study.
We hope, however, that, as in the case of the computer
study, particular aspects of the subject will then be
examined in greater detail at regional and national
meetings.

---

† University of Southampton (D Schonland), University of
  Paris-Sud at Orsay (E de Turckheim), Eötvös Lorand
  University and other institutions in Budapest (T Nemetz),
  Jadavpur University, Calcultta (D K Sinha), Eindhoven
  Technical University (F H Simons), Florida Agricultural
  and Mechanical University (D Hill), and University College
  Cardiff (C A Taylor).

‡ Professor A G Howson, Faculty of Mathematical Studies, The
  University, Southampton, SO9 5NH, England. Professor J-P
  Kahane, Mathématique, Bâtiment 425, Université de Paris-
  Sud, Centre d'Orsay, 91405, Orsay, Cédex, France.

THE QUESTIONS

Although it might not have universal acceptance, we shall take it as axiomatic that mathematics is taught as a service subject in response to a <u>need</u> (depending, naturally, on the major discipline concerned). What need? And what content and methods does this suggest? We propose to reflect on the three questions which arise (Why? What? How?) in the light of what might be done, of positive experiences encountered, and of open problems, rather than provide a simple description of the current state of affairs.

1 WHY?

Why do we teach mathematics to the students of discipline X? There is no generally accepted answer to such a question. Of course, the responses will depend upon the particular discipline X, but we are also likely to obtain different responses from the specialists in X, from their students, and from the future employers of these students - each will hold different opinions.

1.1 In what way will mathematics be used in discipline X? One example of a possible response is given by consideration of the award of the 1985 Nobel Prize for Chemistry to the two mathematicians, H H Hauptman and J Karle, for their development of methods, based on Fourier analysis and probability, for determining crystal structures†.

In Physics, historical examples abound (Mechanics, Relativity, Quantum Theory). Currently, recourse to simulation on a computer has once again brought together physicists and mathematicians, and has given new impetus to some mathematics so far little known (fractals). Informatics (computer science) could not be understood without mathematics, and the recent development of finite mathematics has been a direct response to its needs. Now Chemistry is beginning to rival Physics and Informatics as a valuable source of varied mathematical problems - as has just been shown by the award of the Nobel Prize to the crystallographers. In Medicine, specialists make use of sophisticated tools which necessitate interaction between them, physicists and engineers; mathematicians should play a part in the training which is needed. Biology and Economics are great users of statistical models. Linguistics,

---

† In W Lipscomb's words, "The Nobel Prize for Chemistry is all about changing the field of chemistry. And this work changed the field."

Geography and Geology have developed concepts and techniques which are made more readily accessible by a good mathematical understanding. Engineers, in all their branches of activity, have to calculate, to test hypotheses, and to construct models; must they be restricted in this to the use of traditional tools? On the other hand, is it possible for them to be acquainted with all the mathematics which could prove of use to them in their professional life? Recent events have shown that not all the mathematics which can be applied is to be found within that area conventionally called 'applied mathematics' (for example, algebra and the theory of numbers have been utilised in coding theory and cryptography, algebraic topology in the chemistry of large molecules).

1.2 Since our teaching cannot encompass all the mathematics which might conceivably be used, what then are to be the criteria for selection?

1.2.1 <u>First approach</u>: the student must be capable of making use of those tools with which he† is provided. He must therefore be restricted to concrete questions, techniques and concepts. The best motivation is supplied by considering examples drawn from his own discipline which can be solved using those mathematical techniques and concepts to which he has been introduced. He must shun abstract notions not immediately tied to applications.

1.2.2 <u>Second approach</u>: the student has at his disposal computers and software. This disposes of the need to teach many traditional techniques and skills, but creates a demand for other qualities. The student must know where to turn for help, what he can ask of the computer, and how to guide and control the machine. He must develop the knowledge and skills required to do this. The part mathematics will play in this is as a mode of thought, a mental exercise, and an apprenticeship in rigour.

1.2.3 <u>Third approach</u>: the student has less need to <u>do</u> mathematics than to know how to read it. The professional literature is what will sustain his continuing development, much of it making use of mathematics. He must therefore be taught to study mathematics as a language rather than as a

---

† For linguistic simplicity our typical student will be male. Nevertheless, we hope that this will not be interpreted as sexist. Certainly in all countries there would seem to be a great need to increase the percentage of females studying those subjects which make heavy mathematical demands.

tool. He must be taught how to read it, to consult and use
references. Mathematics assumes its important position as
an element of culture and as a constantly developing
science.

1.3 These three approaches lead, naturally, to different
choices of content and teaching methods. We will return to
this in later sections. Let us begin, however, with three
opinions regarding why mathematics is taught to students of
another discipline.
      First opinion (expressed by students in economics at
Budapest): the only justification for teaching mathematics
is that it weeds out the bad students, because of the
obstacle the mathematics examination presents.
      Second opinion (expressed by mathematicians at Orsay):
a justification for this teaching is that it teaches
students how to use mathematics correctly and to
distinguish, for example, how to construct a suitable model
and to use the mathematical techniques associated with that
model.
      Third opinion (expressed by biologists at Orsay): it
doesn't matter what mathematics is taught, if it is good
mathematics; what is important is that students learn to
reason mathematically.
      Are these opinions completely idiosyncratic - or are
they to be found expressed elsewhere?

2 WHAT

      What mathematics should be taught?

2.1 A variety of very different possibilities arise
depending upon the mathematical knowledge and understanding
which students have gained at school. In some countries it
may even be the case that students have opted out of school
mathematics courses, and then find at university that their
chosen subject, eg Biology, can have a considerable
mathematical component. In certain cases, the initial goal
of universities appears to be to bring all students to a
common level through the teaching of basic techniques
already met - but possibly not learned - at school. Where
this goal is attained it raises questions concerning
previous failures at the school level. Where failures occur
the consequences are dramatic both for students and
institutions (for example, in Florida, before they are
allowed to enter the third year of a state university all
students must pass a 'low-level' test in language and
communication skills which depresses the standard of
mathematics taught). At the other extreme, students enter
university with a strong mathematical background, and are as

well equipped to tackle new and demanding mathematics as
those who have opted to become mathematicians (this is the
case of many engineering students at Jadavpur University and
of those entering the École Supérieure d'Électricité at
Orsay)†.

2.2 Current practice would appear to depend considerably
upon national traditions. Thus at Southampton, second-year
Physics students are taught partial differential equations,
numerical analysis, tensors and finite group theory, none of
which is taught at that stage to students at Orsay.
However, third-year students at the latter institution meet
Lebesgue integration, Hilbert spaces and Schwartz
distributions, subjects not taught at Southampton (but in
the syllabus at Eötvös Lorand University, Budapest).How is
one to explain such differences, and are they as
irreconcilable as, at first sight, they appear?

2.3 We must draw attention here to two specific constraints
on service teaching: the limited time available, and the
fact that many students lack motivation. The former forces
us to accept as axiomatic that service teaching can never
supply students with all the mathematics they are likely to
need.

2.4 Faced with these constraints the universities at
Southampton and Orsay have adopted different attitudes.

2.4.1 First attitude: the primary purpose of mathematics
service teaching is to acquaint the students with the
mathematical techniques that will be useful or essential to
them in their other courses and to give them some confidence
in handling these techniques.

2.4.2 Second attitude: it is a matter of not elaborating and
of moving quickly; for this one must emphasise modern and
powerful tools and be prepared to forget about those tools
whose life is limited - even if they are immediately usable
in other courses.
        In practice, things are not so clearcut. The
Southampton report gives, as a secondary objective, the need
to give students an idea of the scope and power of
mathematics, and to add to a 'utilitarian' approach certain

---

† That such students could follow any mathematics course
  reinforces the need to ask 'Why?' and 'What?' on their
  behalf. Although it lies outside the scope of this study,
  it is, of course, still essential continuously to pose the
  questions 'Why?' and 'What?' and 'How?' in relation to all
  undergraduate courses in mathematics.

'cultural' overtones. At Orsay there is an insistence on
the negotiation of programmes between mathematicians and
other subject specialists - it is not sufficient to travel
quickly, there must be agreement on the general direction.

2.5 The question of what one should teach gives rise to
greater problems since it is inseparable from the questions
'who decides?' and 'who teaches?'

2.5.1 The logic of the first attitude is that, as far as
possible, it should be the teachers of the major discipline
who teach the mathematical concepts which they will then
use. They are aware of the needs, and the introduction of
the mathematical ideas can be timed immediately to precede
their application. This is the situation realised in
Physics teaching at Cardiff and in Economics at the Karl
Marx University, Budapest. The advantages are obvious: for
coherence in teaching, motivation of students and a uniform
use of language and symbolism†. In fact the teachers' aims
go beyond the utilitarian; for the physicists at Cardiff the
mathematics must "help in the understanding of physical
concepts and in the interpretation of experimental results"
- criteria which have a fine ring, are all-embracing and are
operable in all service teaching and do not exclude the
cooperation of mathematicians. The engineers at Cardiff,
however, see things somewhat differently. There the
mathematics courses, jointly agreed and mainly classical,
are given in the main by pure mathematicians, a state of
affairs which the engineers do not find entirely
satisfactory: "Engineering students should be taught by
engineers, or at least by mathematicians who are based in
the Engineering Faculty. The biggest single problem is
motivation, and this is best achieved if the teaching is
done by engineers who are respected by the students as
engineers and who can draw examples to illustrate the
mathematics from their own work. ... Mathematics for
engineers must be taught as a means to an end and not as an
intellectual discipline for its own sake and it is difficult
for mathematicians to come to terms with this."

2.5.2 The logic of the second attitude is to place
responsibility in the hands of the mathematicians (the
case,say, at Jadavpur). It is a question initially of
identifying the needs of the major discipline. Following

---

† An interesting consequence of this policy at Cardiff is
  that physicists are not specifically examined in
  mathematics: motivation for studying mathematics is
  intended to be gained from its teaching being so closely
  bound up with that of the physics.

this the goal will be to model "non-mathematical situations in mathematical terms which, apart from ensuring better insight into the situation involved, enables one to acquire a grip on problem-solving" and "to give a quantitative framework ... a rational and scientific base". In every case, according to Jadavpur University, the mathematician must acquire the language of the [other] discipline, adapt it to a mathematical framework, provide a mathematical analysis, and then translate the results back into the user's language. Such a process, which is most ambitious and demands extremely strong interactions, is to be found at the research level between mathematicians and workers in other disciplines. Even though its realisation at a service teaching level might only be partial, it will have the advantage of permitting the mathematician to construct a coherent course with clearly identified goals. The duty of the mathematician is to construct the most straightforward and shortest course likely to attain these goals - in effect, what he is called upon to do in any course he gives. This might call for a wide knowledge of mathematics.

2.5.3 The two approaches are, in fact, compatible. Here, for example, we can quote a brave proposition advanced by E Roubine (École Supérieure d'Électricité) for the education of Engineers. "Long term aims make it inevitable that there should be a break between mathematics and other teaching. It is reasonable to envisage a foundation course, relatively short, modern and at a high level, essentially of functional analysis (being built, today, upon numerical analysis). In other teaching one can devote a few lessons to reviewing other appropriate mathematics with the symbolism and language best suited to the immediate demands. Well carried out, this could suffice for the entire course." Thus algebra would naturally precede a course in computer science, statistics and probability those in agriculture, and coding theory one in telecommunications.

2.6 A strong argument for an initial mathematical education at a high level dissociated from immediate applications, is the power of computers. They demand that the user should become familiar with ever more sophisticated theories, for as Roubine demonstrates they now make available as everyday tools what were previously theories with little practical application. Thus, for example, Poincaré attempted to apply Fredholm theory of integral equations to aerials. Only, however, in the last ten years have engineers, with the aid of computers, been able to get to grips with singular integral equations.

2.7 Mathematical progress, and the revival of some older topics under the influence of the computer, force syllabus revisions. Pressures will also arise because of progress in the other disciplines (for example, the study of such complex phenomena as polymers and imperfect crystals). Here are a few specific questions.

2.7.1 What is the essential basic algebra and analysis which we should like all students to know? What can be acquired at school level? What must wait until university?

2.7.2 What are the 'traditional' subjects which have been given new life by the computer and today's applications? A typical example arises from differential equations. "Special functions" are now scarcely taught to mathematicians, yet one finds them in the syllabus for chemistry students at Jadavpur. Does the role of symmetry in Physics and Chemistry suggest a place for 'classical groups and special functions'?

2.7.3 What geometry should be included? (The geologists at Budapest still hold on to traditional elementary geometry and descriptive geometry. Solid-state physicists and chemists are interested in polyhedra. Everywhere there are demands for geometric interpretations. Is there a case for introducing fractals and the corresponding mathematics (Weierstrass, Cantor, von Koch, Hausdorff ...)?).

2.7.4 What is the place of statistics and probability? Should these be introduced piecemeal as needs arise, or presented as a structured course? The response may differ in, say, Physics, Biology and Economics. There have also been interesting experiments over some years in medical education.

2.7.5 What is the appropriate mathematics for computer scientists and who should teach it? Wouldn't its algebra, algorithmics and finite mathematics be equally appropriate for other students?

2.7.6 Several institutions now list 'operational research' as part of the mathematics syllabus. How should this be interpreted? Is OR, in fact, a part of mathematics or rather an independent (as yet minor) discipline which should itself be seen as being served by mathematics.

2.7.7 Extreme positions are expressed on certain topics for engineers, for example, Schwartz distributions: useless? indispensable?

2.7.8 Is the teaching of mathematical modelling - a 'necessity' (Jadavpur) or 'a beautiful dream' (Budapest)?

3 HOW?

In the best possible way. And it could be argued that
once it has been decided what should be taught and who
should teach it, then it is a matter to be determined solely
by the individuals concerned. There are, however, many
general points which merit particular consideration.

3.1 Statements and Proofs

There can be no justification for giving statements
which are incorrect, for example, for stating - or
suggesting - that the Fourier series of a continuous
function converges uniformly to that function. Yet there
will be times when the teacher wishes to make statements
because they are simple and correct in a convenient frame.
For example, an integrable function on R tends to zero at
infinity (in the sense of distributions). Each function on
R is Lebesgue-measurable (in a model of set-theory which
excludes the axiom of choice). Each part of a probability
space is an event (in the same model). An essential point
is to make useful statements in the most primitive possible
language.

The choice of good definitions and statements is the
work of a mathematician, but one in which non-mathematicians
can usefully participate. It must also be recognised that
there is nothing sacrosanct about the order in which
material is presented. For example, it is not forbidden to
define the rotation (curl) of a vector field starting out
from a physical interpretation of Stokes' Theorem (Berkeley
Physics Course), rather than from the usual operator
definition in terms of derivatives: the theorem can precede
the definition or vice-versa.

In a course given to mathematicians the guarantee of
exactitude and of cohesion is the chain of logical argument,
proof. In a service course then sometimes one must replace
proof (too long, non-illuminating) by other arguments, and
develop, for example, what George Polyá termed 'plausible
reasoning'. Good physical illustrations can be more
enlightening and impressive than proofs: depressing the
sustaining key on a piano, saying 'ohh' to the strings, and
hearing the response 'ohhhh' is an excellent gateway to
spectral analysis and synthesis (Berkeley, Waves, p 91).

On the other hand, exploratory work and verification on
a computer can give certain mathematical statements the
status of 'experimental' truths. Mathematical rigour
consists in distinguishing between mathematical proof and
experimental verification - this distinction must not become
blurred.

3.2 Examples and concepts

Must one begin with examples and from these derive the concepts, or should one start off with the concepts and flesh these out with examples? This is an old question. Should one restrict oneself to examples drawn from the major discipline? Advice varies and depends upon many external constraints, in particular, the time available and class size.

One possibility merits special attention: this is the introduction of exploratory data analysis at the beginning of university studies. Manipulation can be done without any great theoretical apparatus; in addition, important motivation can be provided for the study of linear algebra and probability.

In general, the relations between examples, concepts and intuition generate major pedagogical questions. The great unifying concepts (groups, measure) are not accessible, despite their apparent simplicity, unless they are supported with numerous illustrations and examples. This is true of all mathematics teaching, but unfortunately within service teaching students are not provided with the time in which such notions can become familiar and intuitive.

3.3 Small or large classes?

Generally, the response to this question depends almost entirely on local resources: large groups demand fewer teachers. There are clear administrative advantages in the tradition of teaching service mathematics to large groups, often drawn from several different departments: economies of preparation of both lectures and exercises, and the possibility of employing only such lecturers as have a direct interest in service teaching and who, over the years, amass experience concerning likely points of difficulty, general needs etc. The disadvantages include the lack of motivation for the students, the restrictions placed on the kind of learning activities which can be offered, and the impossibility of setting a common examination which matches the real needs and strengths of students drawn from a range of departments. (We note, however, that at Eindhoven, even though the 'large group' format has been retained, this has not prevented the introduction of a novel course which depends upon each student having his own programmable pocket computer).

The question is also bound to that of 'who teaches?'. The case for having a large inhomogeneous class taught by a mathematician is very strong. Small groups, on the other hand, are better able to utilise exercises and examples which draw on their major disciplines.

Even with any one discipline, however, first-year students are likely to differ very greatly in their mathematical attainments and abilities. This creates difficulties for the lecturer and forces consideration of other methods of teaching and learning. For such reasons, we should like our study to pay particular attention to experiments which have been made to help resolve such pedagogical problems. We note, for example, that at Southampton first-year engineers follow an individualised, 'self-paced' course based on reading (with frequent testing) rather than lectures.

3.4 The 'Ideal' situation
Subject to the various constraints which have to be met, what patterns of service teaching are giving rise to local satisfaction? We have already referred (Section 2.5.1 above), to the way in which mathematics is taught to physicists at Cardiff. Here we give other examples of situations considered 'ideal'.

At Southampton, the course for chemists is given by a mathematician but each student, together with three or four others, is seen fortnightly be a chemist who will give tutorial supervision using material and example sheets supplied by the mathematician†. A similar system has operated for some years in the Physics department at Orsay to general satisfaction: the lectures to the whole class being given by a mathematician, directed work (to groups of 20 students) by physicists.

At Paris-Grignon, a 20-hour course for third-year students of Agriculture was mounted in the form of a dialogue between an economist and a mathematician, thus providing the framework for an effective investigation. Such 'team-teaching' is very motivating for students, but is very expensive in preparation time.

No doubt other 'ideal' situations having different characteristics can be found. Detailed descriptions of them would be extremely welcome.

3.5 The use of computers
As was written above, the impact of computers on the teaching of mathematics has already been the subject of an

---

† One chemist wrote of this arrangement: 'many of my colleagues agree with me that in Southampton Chemistry we have the ideal situation as far as academic considerations are concerned. In tutorials the chemists can relate the material covered to Chemistry, point out the relevance to the Chemistry course and (it is hoped) provide some motivation'.

ICMI study. It is essential that we reflect on all the new
possibilities offered by computers (rapid computation,
graphics, experimentation) and on the changing needs caused
by their introduction (changes both of curricular content
and also of desirable qualities to be developed in
students).
    A feature of the reports we received was the limited
use of computers in the teaching of those subjects which
have traditionally made heavy use of mathematics.

3.6 **The use of books and papers**
    Here there are two aspects. First, for service
teaching, it is good to use texts written collaboratively by
mathematicians and specialists in the major disciplines.
Such books do exist, but there are many gaps. It would be
valuable to have the characteristics of the successful
texts, and also the lacunae, described.
    Secondly, as we have already stressed, students must
learn how to read mathematics, both in order to learn more
mathematics when there is no lecturer to hand, and also to
understand their professional literature. Descriptions of
'planned' reading tasks are not numerous, but appear of
interest and potential value (eg readings of extracts from
Laplace for students at the Paris École des Ponts et
Chaussées, a chapter of Volterra for biologists at Orsay).

3.7 **Examinations, assessment and control**
    In many cases examinations supply the principal
motivation for students (although, as we have indicated in
Section 2.5.1, this need not necessarily be the case). If
the examination is outside the lecturer's control (as in
Florida, and even more in the preparatory classes for the
'grandes écoles' in France), then it also provides
motivation for him. Therefore, the questions 'Why?' and
'How?' should not be asked of teaching alone, but must also
be asked of evaluation and assessment. If the teaching of
mathematical modelling is a primary goal, then this goal is
unlikely to be attained, if all that is required to pass the
examination is memory of a ragbag of techniques applied in
stock, purely mathematical situations. On the whole
examinations tend to freeze courses, and militate against
such innovations as, for example, the introduction of
computers, mathematical modelling, and 'planned' reading.
On the other hand, all of these innovations can be
effectively examined, and examples can be given. However,
their assessment is extremely time-consuming and the large
numbers of students involved in service courses present
particular difficulties.

How, then are we to use examinations and assessment as
a means for <u>improving</u> teaching and learning? What desirable
changes can be made to entrance examinations or to national
examinations? Are there forms of continuous assessment
which enable teachers/students to monitor the assimilation
of the mathematics they teach/learn? Can this be done
within the short time allocated to service teaching? Are
there still examinations which contribute little and might
be better abandoned? Examples of good practice will be
welcomed.

4.   CALL FOR PAPERS

In this discussion document it has been possible only
briefly to indicate some questions of great interest and
concern. The next step is to take a selection of these and
to delve into them more deeply, to flesh arguments out with
examples taken from current practice, to examine
philosophical and pedagogical points more critically, to
report the results of relevant research. The Planning
Committee for the study would very much welcome papers which
so develop points made in this paper, and which, in their
turn, could form the bases of discussions in Udine in April,
1987. Such papers would be welcomed from all concerned with
service teaching, mathematicians, specialists in other
disciplines, students, recent students and employers.

TEACHING SERVICE MATHEMATICS: REMARKS FROM A THIRD WORLD PERSPECTIVE

I.A. Ahmed
University of Gezira, Wad Medani, Sudan

ABSTRACT

The teaching of mathematics as a service subject raises problems in any context. This paper presents the observations of a teacher of service mathematics in the Sudan and provides a Third World perspective on those problems.

1. INTRODUCTION

My remarks derive from my experience of teaching
"service mathematics" during the years 1970-1986, the first
eight of which were spent in the University of Khartoum and
the last eight in the University of Gezira, Sudan. During
these years I have taught service mathematics, both at the
undergraduate and postgraduate levels, to students of a wide
range of subjects including science, engineering, geology,
economics, medicine and agriculture.

I shall concentrate mainly on "why" and "how" we teach
service mathematics rather than on "what" we teach. This is
because answering the two questions "why?" and "how?" raises
issues concerning ends and means or methodology which are
central to any attempt to organise and bring order to
education in any field of learning. In the case of
mathematics, whether as a discipline in itself or as a
service subject, answering these two questions assumes extra
relevance and urgency due to the all-pervasive nature and
widespread uses of mathematics. This concentration on "why"
and "how" should not, however, be taken to imply that
content (the question of "what" to teach) is of minor
importance.

After dealing with why we teach and learn service
mathematics in section 2, section 3 relates some positive
experiences and offers suggestions on how to teach service
mathematics. This is followed by section 4 which touches on
some structural aspects of "what" to teach. The paper ends
with some brief conclusions in section 5.

2. WHY?

As Head of the Department of Mathematics, Applied
Statistics and Computer Science at Gezira, I sent out a
circular to all users of service mathematics throughout the
University. The considerations which motivated this
circular were many and varied, ranging from the
philosophical through the educational to the down-to-earth
practical, and were dictated by the scarce and meagre
resources at our disposal. The intention was to encourage
users to reflect on their needs and pinpoint their aims and
objectives in teaching and learning mathematics.

The response to the circular was very discouraging.
Few responded and even those who did do so did not appear to
be clear in their own minds about their real needs and
objectives for service mathematics. All they knew was that
they needed mathematics because they vaguely sensed its
importance to their respective disciplines and because they
did not want to be left out or behind in this age of the

computer which they identified with mathematics in an ill-understood manner. Worse still, some, out of inertia of habit, wanted to see service mathematics taught because that had always been the case.

The fear of being left behind, or equivalently the desire to keep abreast of new developments and innovations regardless of whether these innovations meet the requirements of relevance and appropriateness or not, is indeed one of the worst reasons advanced as an answer to the question: Why should we teach or learn service mathematics? Yet this kind of attitude is very widespread, in various contexts, in the Third World, resulting in the adoption of all sorts of inappropriate technologies, the introduction of irrelevant studies and the proliferation of the latest gadgets which are rendered obsolete the day they come out of their wrappers. Needless to say, such gadgets are usually poorly used, badly maintained and ill-repaired, if at all, when they cease to function. This comment becomes even more relevant and appropriate when we bear in mind the rash and mindless call for the computerisation of all departments of mathematics in the Third World as part of a basic strategy for "catching up" with the First World.

Returning to the response to the circular described above, the most striking conclusion is that users of mathematics are seldom very clear in their minds about why they want to learn or teach mathematics. Hence, it seems to be imperative that mathematicians should guide their colleagues from other disciplines in defining their objects and aims. This must not be interpreted by users as paternalism or academic chauvinism on the part of mathematicians. To allay the fears which might be engendered by this situation, mathematicians need to collaborate with the users of service mathematics in multi-disciplinary groups. This approach, based on teams interacting in an inter-disciplinary manner, has also the additional value of exposing mathematicians to a first hand taste of the needs of users, helping in the joint clarification and identification of objectives and facilitating agreement on suitable and appropriate methods for realising those objectives.

My previous experience at the University of Khartoum, teaching service mathematics to first year biological stream students, corroborates the perception of service mathematics noted in [1], that it often serves as a filter to weed out the weaker students because of the obstacle the mathematics examination presents.

Having described some experiences and the lessons drawn from them I end this section with a quotation from the circular referred to above:

"Serving as an academic lingua franca, mathematics, in the widest possible sense of the word, is well-poised to bridge the gulfs that seem to separate all fields of human knowledge, with the concomitant fragmentation, schisms and divisions which militate against the spirit of universality and unity, sorely needed by our turbulent age. A hopeful sign, in this direction, is the massive mathematisation process now underway in all fields of human knowledge."

This role for mathematics, as an academic lingua franca, is my partial answer to the question of why we should teach service mathematics. The other part relates to the cultural and intellectual enrichment which studying mathematics bestows on its students and users. Clearly this is a variant of the themes expressed in [1], and again indicates that those opinions are not idiosyncratic but are found elsewhere.

3. HOW?

The question of how to teach service mathematics is vital and the success of any programme of service mathematics critically hinges upon it. Experience has shown that any answer to this question is always unsatisfactory in some sense. As a result, for much of the time, students of service mathematics lack motivation and a sense of purpose.

For this reason I have always attempted to create a sense of purpose and direction by relating the topic being treated to its historical roots and pointing, as it were, to the peaks above as urged by Thwaites [2]. Relating a topic to its roots helps in tracing the genesis of ideas to situations concrete enough to inspire confidence and inject a sense of meaning and purpose, while pointing to the peaks above opens a window on the future, thereby creating a sense of direction and a desire to march forward. Indeed it is my opinion that, to deal successfully with any present situation, it is essential to consider it as the dialectical synthesis of the future and the past, that is the present moment must be viewed as a Janus-face representing the confluence and interaction of forces proceeding from the past and forces advancing from the future (see Taha [3]). A classical example that lends itself to the above treatment is calculus and the same applies to plane geometry, analytical geometry and many other fields.

Another approach which turned out to be very effective, after encountering some initial resistance, is insistence on exact statements. Experience has shown that even the most fundamental relations such as equality are poorly understood, and many otherwise able students fail to balance

the two sides of an equality which betrays a fundamental
lack of understanding of mathematical logic and thinking.
Hence, by monotonously insisting on correct equality
statements, students invariably begin to develop the
required mathematical understanding.

It is also essential that every effort be made to teach
service mathematics with a view to enabling students to
develop an integrated attitude to mathematics and to become
trained in and well acquainted with mathematical ways of
thought.

To conclude this section on "how" to teach service
mathematics, I will briefly consider the service mathematics
teacher. Teaching is an art which many university teachers
never find the time or opportunity to perfect. All sorts of
pressures, real and imagined, tend to divert them from this
desirable end. That is why we must

"allow those who like teaching to teach. We should
not create such a situation that one feels compelled
to do research." (Lee [4])

We must also resist external pressures generated by the
search for recognition by foreign peers. This is a trap
which leads to alienation and to making improper,
inappropriate and irrelevant choices. To avoid this it
might be useful to heed the recommendations:

"At all levels of education, educators in developing
countries should be encouraged to pay less attention
to the apparent demands of academic respectability, as
judged by their professional counterparts elsewhere in
the world, and more to the alignment of their work
with the genuine needs of the people whom they and
their institutions serve." (Howson [5])

and also

"When experts are consulted for promotion of young
mathematicians, they could be asked to base their
judgement and advice not only on the research papers,
but also on the general contribution to teaching
and/or the implementation of applications." (Lion [6])

Indeed, we must insist on continuous education and
endless retraining as a strategy for raising the standards
of teaching and education. As a motivation, this could be
used as a basis for promotion. According to Ahmed [7]

"Instead of demanding research papers as a basis for
promotion, we could demand continuous education of
teachers so as to raise personal and general
standards. This continuous education could be
achieved through planning continuous training in
advanced and important topics for both old and new
staff members. To facilitate this, staff members
could be awarded sabbatical leaves, short training

periods and study visits to advanced research centres, curriculum development centres and interdisciplinary centres which seek to solve problems in the different fields of knowledge."

4. WHAT?

As mentioned in the introduction, this section will deal with the structural aspects of "what" service mathematics to teach. In this respect it is sufficient to state the following broad guidelines :-
a) Due to the fact that knowledge and skills are rapidly changing, it is "desirable to impart invariant knowledge to the pupils and have them develop invariant skills." (Van Weert in Lovis and Tagg [8])
b) "Emphasis must be placed on that part of mathematics which depends on visual imagination: the drawing and interpretation of graphs, the recognition of curves, the appreciation of pattern, the visualisation of motion and of three-dimensional relationships." (Howson [5])
c) Part of what is taught must be concerned with teaching the students to write the language in which instruction is carried out clearly and grammatically.

5. CONCLUSIONS

The following are some conclusions which may be derived from this paper :-
1) Users of service mathematics are seldom very clear in their minds about why they want to learn or teach mathematics. Hence it seems imperative that mathematicians should guide their colleagues from other disciplines in defining objectives and aims. However, this must be done in a collaborative fashion through multi-disciplinary teams working in an inter-disciplinary fashion.
2) Teachers must be allowed to teach by screening them from unnecessary pressures, encouraging them and offering them every opportunity of continuous education and training.
3) The content of a service course should be invariant knowledge rather than transiently fashionable knowledge, emphasis must be placed on that part of mathematics which depends on visual imagination and students must be enabled to master the language of instruction.

REFERENCES

1. A.G.Howson et al, <u>Bulletin of the Institute of Mathematics and Its Applications</u>, 22, 1986, pp 85-90.
2. B.Thwaites, <u>On Teaching Mathematics</u>, Pergamon, 1961.
3. M.M.Taha, <u>The Cultural Revolution</u> (translated by I.A.Ahmed), Deena, Khartoum, 1971.
4. P-Y.Lee, Development of mathematics in South-East Asia: the experience of the South East Asian Mathematical Society, in M.E.A.El Tom (editor), <u>Developing Mathematics in the Third World</u>, North-Holland, 1979.
5. A.G.Howson (editor), <u>Developments in Mathematical Education</u>, Cambridge University Press, 1973.
6. J.L.Lion, Computers, mathematics and applications, in M.E.A.El Tom (editor), <u>Developing Mathematics in Third World Countries</u>, North-Holland, 1979.
7. I.A.Ahmed, <u>The future of mathematics in the Arab world</u> (in arabic), paper presented at the Symposium on the Relation Between Science, Technology and Society, Doha, Qatar, 1-4 December, 1986.
8. F.B.Lovis and E.D.Tagg (editors), <u>Informatics and Teacher Training</u>, Elsevier (North-Holland), 1985.

TEACHING MATHEMATICS TO ENGINEERS: SOME REMARKS ON THE ITALIAN CASE

A.Bacciotti, P. Boieri
Politecnico di Torino, Torino, Italy

ABSTRACT

In this paper the structure of mathematical teaching in Italian school system is briefly outlined, emphasizing the role of Mathematics in Engineering Faculties.
Some data about mathematical background of students entering University and percentages of students enrolled that continue and graduate in Engineering are provided.
The distribution of Mathematics teachers in the different Faculties is shown and some common attitudes towards teaching Mathematics as a service subject are examined. The relation between Mathematics and applications, the necessity of rigour and abstraction are stressed.
In the final part, an example of integration between mathematical theory and applied modelling is shown.

This paper is the authors' contribution to the discussion about teaching Mathematics to non Mathematics students arisen by the ICMI document "Mathematics as a service subject" ([1]).

The following remarks are based on the authors' teaching experience. Hence, they mainly refer to teaching Mathematics in Engineering faculties. Some teaching experiments that can support and complete our general statements are also presented.

Since the reader is not supposed to be familiar with the organization and structure of Italian school and university systems, some preliminary remarks are needed.

1) Children enter the Italian school system when they are six years old. Primary school is eight years long and it is compulsory. Secondary school is five years long and it is divided in several branches: letters, sciences, technical schools etc.

According to a research of ISTAT (the central institute for statistics of Italian government), reproduced in [2], students attending secondary schools in 1984 were 56.4% of the corresponding age cohort. In the same year, the number of students who entered University was 65.7% of the number of students who finished secondary schools.

It is important to notice that students enter University without examinations and regardless of the branch of secondary school they attended.

Mathematical topics usually taught in secondary schools are elementary algebra, euclidean and analytical geometry, trigonometry, logarithms. In the scientific branch also elements of Calculus are introduced.

2) The degree in Engineering is conferred at the end of a five years' study: the first two years are devoted to the "service" subjects, while the last three focus on the specialized Engineering curriculum. With a few exceptions, the mathematical teaching is concentrated in the first two years and therefore separated from the specific Engineering teaching.

3) Concerning the students mathematical background, the outcomes of an investigation made a few years ago at Pisa and Genova Universities are very interesting (see [3]). First year Mathematics, Computer Science, Engineering, Physics and Biology students were required to answer eight mathematical questions, during the first classroom lesson. These questions were inspired by the content of a Syllabus established by U.M.I. (the Italian Mathematical Society) and covered the (supposed) common background (see [4]). The following tables show the average number of correct answers (out of eight).

Summing up, students enter University with a poor mathematical background and without entrance examination.

For this reason the first courses must proceed slowly. This is compensated by a rather long time devoted to mathematical courses, compared to other countries: for instance, Engineering students attend four mathematical courses, with 80 hours of lessons and 60 hours of problem solving classes each, organized on a whole year or a semester base.

CURRICULUM CHOSEN		HIGH SCHOOL ATTENDED	
Mathematics or			
Computer Sciences	4.7	Scientific	5
Physics	5.9	Letter	4.2
Engineering	4.8	Technical	4.5
Biology	3.1	Others	3.7
Total	4.6	Total	4.6

4) Courses are usually very crowded and they are structured on the base of classroom lessons. Topics taught to Engineering students are Calculus, Linear Algebra, Analytic Geometry, Classical Mechanics, Ordinary Differential Equations. In some courses also introductory Functional Analysis, Complex Variables, Partial Differential Equations and Numerical Methods are covered.

Students are encouraged to make use of computers in studying. For instance, at the Polytechnic of Turin, there is computer laboratory with several dozens of PC's and terminals for first and second year students. Electrical Engineering students must attend a course in basic Computer Science in the first year.

5) The following data concerning Engineering students at Polytechnic of Turin may be interesting. Although national data are not available, we believe that the situation of other Italian Engineering Faculties is not different. The table shows the number of students who, enrolled in the academic year 1980/81 (first graph) and 1981/82 (second graph), "survived" in the successive years and eventually graduated.

The increase of students in the fifth year is due only to a technical fact. Indeed, students who attended all the curriculum courses but did not pass some examinations are enrolled again as fifth year students.

6) According to a recent investigation made by U.M.I. (see [5]), 20.5% of Italian mathematicians with a permanent position in 1983 were teaching courses of Engineering Faculties and 18.6% were teaching courses of Chemistry, Biology, Architecture, Economics, Agricolture and other curricula. Hence, The majority (60.9%) were teaching courses for Mathematics, Physics and Computer Science students (Since in Italy curricula in Mathematics, Physics and Computer Science are offered by the same Faculty, is very difficult to obtain more precise percentages). Today the percentages shown above may be slightly changed, because of the recent opening of new positions. However we believe that they give a realistic idea of the situation.

There are no part-time positions in Italian University.

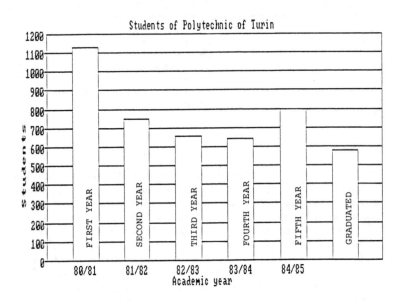

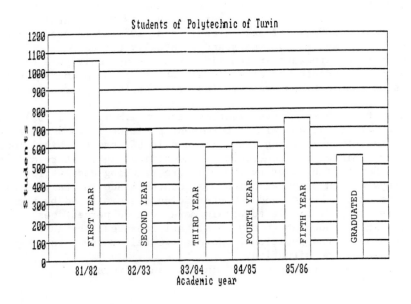

7) Teachers of mathematical courses usually do their research work in the specific Mathematics field; their academic career is based on nationwide competitions where the board of examiners (also selected on a nationwide base) is composed by mathematicians only.

8) Because of lack of constant cooperation between mathematicians and engineers in research, mathematicians usually have a secondary importance in Engineering schools.

9) Continuing education programmes are not pursued in a systematic way in Italy and they usually do not involve University staff.

This situation is far from being ideal. It is a consequence of the traditional Italian academic structure (not earlier than 15-20 years ago Engineering, Physics and Mathematics students attended common Mathematics courses) and affects in a direct way the behaviour of teachers and students.

Concerning the way mathematicians feel about their teaching, we find, roughly speaking, three possible behaviours: lack of a serious interest (lessons based on a traditional textbook, aiming at training students in calculus techniques more than at communicating real mathematical theories and methods), engagement in a traditional style (a quite deep and careful way of teaching, but based almost only on theory rather than on applications), "illuministic" type of work (constant and personal effort to combine serious mathematical contents with examples and applications to Engineering).

From the students' point of view, the lack of integration between Mathematics and applications rises doubts about the reason why some topics in Mathematics are taught. The less mathematically minded among them study mainly for the necessity of passing a very selective examination.

Also, the engineers who teach the specialized topics are sometimes scarcely acquainted with the contents of mathematical courses and the progress in mathematical education. Sometimes we feel that the necessity for a deep mathematical background for an Engineering student is a matter of doubt.
More precisely, everybody agrees that Engineering, Economics and Life Sciences require very advanced mathemathics on a research level. However, it is a rather widespread opinion that most Engineering graduates will find routine jobs, where they will be required only to use formulas discovered and verified by others.
Engineering students should be able simply to "read" someone else's Mathematics rather than to understand and apply it.

In these conditions, it is not reasonable to expect substantial changes in the quality of the mathematical contents presently taught.
The reductive interpretation exposed above must be rejected. Indeed,

even if a student will not use in his future job all the Mathematics he
learnt at University, we believe that studying Mathematics is a good
exercise for improving rigour, precision, capability of understanding
and using a formal language; for this reason it should be a necessary
part in everybody's training and, a fortiori, in the training of future
science and technology workers.

On the other hand, we can not teach, in the basic courses, all the
Mathematics applied or applicable to Engineering. Furthermore, a complete integration of the mathematical and technical teaching seems to be
impossible in Italy now, since, as we pointed out above, all the mathematical courses preceed the technical ones.

But also in the present conditions, we believe that the way of teaching can be enhanced to make it possible to train students not only in
the use of traditional mathematical techniques (algebra, derivatives,
integrals and so on), but also in a clear and correct use of logic and
mathematical language.

The necessity for rigour in natural sciences can be proven by emphasizing the role of Mathematics in the process of discovery (from experi-
ment to theoretical models and back, from real situation to abstract
ideas and back). In this way the student is stimulated to take a critical attitude towards a given problem.

A topic particularly suited to be approached in this way is the
theory of Ordinary Differential Equations. They can be introduced in two
ways: as an abstract problem (to find a family of functions that satisfy
some mathematical conditions) and as a model of real problems.

We present now an example from our teaching experience, where we
think that the need for rigour, the connection with real problems (even
if not strictly related to Engineering) and the capability of inducing a
critical attitude are combined.

We consider the dynamics of a population. As a first approach, we are
led to model it with the simplest and non trivial example of Ordinary
Differential Equation:

(1) $$x' = a\,x \qquad\qquad x \in R.$$

This equation can either be solved analytically by elementary methods
or the solutions can be plotted using a numerical method on the monitor
of a computer: in any case, we can study the set of its solutions.

In both cases, we have a theoretical model that enables us to predict
the behaviour of the population as time increases: it is not difficult
to realize that this prediction is not a reasonable one. The student can
be made aware of the fact that, in modelling the phenomenon and getting
equation (1), some essential features have been neglected.

A more careful study leads to the equation:

(2) $$x' = a\,x\,(k-x) \qquad\qquad x \in R.$$

This equation is nonlinear and therefore more difficult than equation

(1). However it can be solved by separation of variables.

At this stage, we can point out the fact that it is not very important to compute the "exact" solution; this solution depends on the
parameters a, k and on the initial value x_0. These constants are derived
from experiments and can be identified only within a certain error
range. For this reason a qualitative study of the solutions is much more
important. Every solution of (2), corresponding to a positive value of
x_0, tends asymptotically to a constant value. This enables us to predict
a saturation phenomenon, i.e. the existence of a stable equilibrium.

This prediction does not always agree with reality: the number of
elements of a species is usually subject to oscillations. This remark
forces us to reconsider the construction of our model and to give up our
attempt of studying the evolution of a population as an isolated entity.
We must take into account the interaction between different species
living in the same environment.

This leads us naturally to introducing systems of Ordinary Differential Equations, starting with the classical model of Lotka and Volterra:

$$(3) \qquad \begin{cases} x' = a\,x - b\,x\,y \\[2mm] y' = -c\,y + d\,x\,y. \end{cases}$$

It is well known that this system has a rest point of center type,
i.e. all the trajectories near the equilibrium are periodical. At this
stage, it is very useful to integrate the system numerically. The deli-
berate use of a not very precise numerical method can emphasize the
structural instability of model (3) and its inadequacy to describe the
real phenomenon.

This research can be continued and can go very far; we think that the
methodological aspects of this way of teaching and its usefulness are
now sufficiently clear, to justify our expectation that its application
can be extended to the whole field of mathematical teaching to Engineers.

As a final remark, it should be pointed out that this teaching method
requires, in several occasions, an intensive use of numerical approxima-
tions, opening up a quite thorough integration between discrete and
continuous Mathematics.

REFERENCES

[1] L'Enseignement Mathematique, 1986, XXXII, (1-2), p. 159
[2] Notiziario dell'Unione Matematica Italiana, 1985 (5), p. 26
[3] Notiziario dell'Unione Matematica Italiana, 1984 (1), p. 28
[4] Notiziario dell'Unione Matematica Italiana, 1980 (3), p. 5
[5] Notiziario dell'Unione Matematica Italiana, 1983 (10), p. 44

TEACHING MATHEMATICS TO ENGINEERS IN WEST GERMANY

R. Böhme
Ruhr-Universität, Bochum, FRG

ABSTRACT

This paper describes the general situation of mathematical service courses, particularly for engineering students, in tertiary education institutions in the Federal Republic of Germany.

1. A REPORT ON THE CURRENT SITUATION

The system of education in the Federal Republic of
Germany in 1987 must be considered an "open" system at all
its levels. As a general rule, success in the final
examination at one level qualifies the student for admission
to the next level, but not necessarily to the students'
preferred institution. As a consequence the system produces
high failure rates. Usually students give up their studies
at a particular level when they realise how much work and
effort are required to succeed at that level. Under this
system, therefore, any student with the final qualification
from the secondary schools (the Abitur) has access to
courses in any discipline of engineering or science, but
perhaps not a free choice between the universities.

The major problem for teachers of mathematics is that
the subjects which are taught before university vary greatly
and do not, even approximately, achieve the same level.
This problem is widely recognised, but the German political
system makes it difficult to solve. Therefore any course of
mathematics in the applied sciences and in engineering is,
in fact, a filter, because all teachers of mathematics must
necessarily assume a certain body of prior knowledge of the
subject. By tradition and convention in Germany, the level
of first courses in mathematics for engineers is almost
constant, and can be regulated by the "Deutscher
Fakultätentag". On this basis there is also a general
consensus about the topics and the style of mathematics for
engineers. Students are assumed to need knowledge of a wide
range of mathematical topics but not necessarily to the same
depth as would be required for mathematics students. All
teachers therefore need to be able to cover this wide range
of topics.

Students have to learn computational methods in order
to be able to obtain definite, numerical results. They are
expected to learn how to attack mathematical problems, how
to use the literature efficiently, and also how to work
together in groups. It is reasonable to hope that a
professor will not present an indirect proof if a
constructive argument is feasible and more or less
equivalent, and that he will present examples of the theory
taught which are chosen from the engineering discipline
which the particular students are studying.

Generally the mathematical topics taught to engineers
are the calculus of one and several variables, linear
algebra, differential equations, Fourier series and special
functions, numerical analysis and probability theory.
Courses on computer science are now offered by many
universities.

The volume of mathematics which an engineering student has to master is 6-8 hours a week for three or four semesters. Generally this period ends with one or two written examinations and an oral examination. In some universities the mathematics department also offers courses for students in the third year, but only rarely are graduate student courses or continuing education course for those who have already entered employment given. Many German universities, however, offer preparatory courses for intending students ("Vorsemester" or "Nulltes Semester") in the two months prior to the beginning of the university term in the Autumn.

The usual teaching system is that one professor (having a full-time job and tenure) teaches four hours a week in a fairly large auditorium with many students. In addition the students must also attend exercise classes and do homework, possibly given by someone else. The homework is a basis for a grade given at the end of the course.

The total volume of service courses differs between technical universities ("Technische Hochschulen") and others ("Universitäten" and "Gesamthochschulen") and may be fairly large or fairly small. In the author's university (Bochum) there are many service courses. There are a large number of courses in computer science. Every year there are first and second year courses in "mathematics for electrical engineers", "mathematics for civil and mechanical engineers" and "mathematics for physicists and geophysicists", and one-year courses in "mathematics for chemists", "mathematics for biologists", "mathematics for geologists and mineralogists", and also some shorter courses in "numerical analysis for engineers". These courses together make up somewhat more than half the teaching load of the mathematics department, though this may be more than the average in Gemany.

Exact statistics were not available, but it is a reasonable estimate that about 30% of school leavers each year are eligible to attend universities. Perhaps 10% of these consider taking a degree in engineering. The failure rate of students of engineering, measured from entry to university studies to graduation with a university diploma, may be estimated to be between 40% and 60%. In recent years the number of entrants to faculties of engineering has been quite large.

LA SITUATION EN FRANCE

P.L. Hennequin
Université Blaise Pascal, Clermont-Ferrand, France

1. LES ETUDES JUSQU'AU BACCALAUREAT

<u>Préélémentaire</u> (3 - 6 ans)

<u>Primaire</u> cinq années (6 - 11 ans). Un seul maître n'ayant souvent que le niveau du baccalauréat en mathématiques.

<u>Collège</u> quatre années (11 - 15 ans). Un professeur de mathématiques enseignant parfois une autre discipline (physique, sciences naturelles, dessin, musique, sports ...) et n'ayant souvent que le niveau du 1er cycle universitaire.

<u>Lycées professionnels</u> trois ou quatre années (14 - 17 ans). Les élèves y rentrent après 2 ou 3 années de collège trop souvent après un échec. De nombreuses sections conduisent actuellement au Certificat d'Aptitude Professionnelle ou au Brevet d'études professionnelles. Une récente réforme a crée des baccalauréats professionnels. Les enseignants de Mathématiques ont le niveau du 1er cycle universitiare et sont formés dans des E.N.N.A. (Ecoles Normales Nationales d'Apprentissage)

<u>Lycées</u> trois années (15 - 18 ans). Un professeur de Mathématiques n'y enseigne que sa discipline. Il a eu cinq ou six années de formation dont 3 ou 4 de formation théorique universitaire sanctionnée par une licence ou une maîtrise. Il a été recruté par un concours CAPES (Certificat d'aptitude au professorat de l'enseignement secondaire) ou Agrégation. Le niveau de ces concours varie beaucoup de cinq en cinq ans.
Les Lycées classiques ou techniques préparent aux divers baccalauréats. En 1986 il y a eu 262 960 reçus, soit 31 % d'une classe d'âge se répartissant ainsi suivant les sections

```
A       (Littéraires)                                        18 %
B       (Economie)                                           19,6 %
C - E (Sciences mathématiques et Physiques)                  15 %
D - D'(Sciences physiques et naturelles)                     17 %
F       (Industriel)                                          9 %
F8      (Médico-social)                                       3 %
G       (Secrétariat - Gestion - Economie)                   19 %
H       (Informatique)                                        0,5 %
F 11-12 (Dessin - Musique)                                    0,2 %
```

Dans chaque section il y a un enseignement de mathématiques et une épreuve au baccalauréat soit à l'écrit, soit à l'oral.

Les programmes sont nationaux, propres à chaque section et révisés environ tous les cinq ans.

2. LES ETUDES APRES LE BACCALAUREAT

Actuellement en France le taux de scolarisation est de 93 % à 16 ans, de 40 % à 19 ans, de 20 % à 21 ans et de 10 % à 23 ans.

Pratiquement tous les bacheliers tentent de poursuivre leurs études au moins pendant un an mais tous les enseignements postérieurs au baccalauréat ne sont pas délivrés par l'Université, contrairement à ce qui se passe dans la plupart des autres pays. Nous distinguerons les trois cycles, chacun durant en principe deux ans

a) Premier Cycle

Il existe principalement quatre voies correspondant à quatre types d'établissements.

(1) L'Université (hors Instituts de technologie (cf (3) ci-dessous))

Elle accueille pour l'inscription en première année 65 % (+) environ des bacheliers qui se répartissent ainsi entre les diverses formations :

```
Droits et Sciences politiques                       16 %
Sciences Economiques - Gestion                       9 %   (*)
Administration Economique et Sociale                 6 %
Lettres - Sciences Humaines                         35 %   (*)
Enseignement du premier degré                        1 %   (*)
Mathématiques Appliquées aux Sciences Sociales       1 %   (*)
Sciences                                            19 %   (*)
Médecine                                             8 %   (*)
Pharmacie                                            3 %
Activités physiques et sportives                     1 %   (*)
```

(+) On retrouve dans ce décompte une bonne partie des élèves des classes préparatoires qui s'inscrivent en général aussi à l'Université par sécurité.

On trouve des enseignements de mathématiques dans les filières repérées par une étoile (*)

Ils sont le plus souvent donnés par des mathématiciens suivant un programme qui laisse une grande liberté à chacun. Dans les disciplines scientifiques, cet enseignement est en général fractionné (algèbre/ analyse/probabilités ...) et répartis en cours et travaux dirigés donnés par des personnes différentes.

Il n'y a en général aucune sélection à l'entrée mais par contre beaucoup d'échecs (près de 50 %) à la fin de la première année. Environ la moitié des universités ont réorganisé leurs premiers cycles pour réduire les taux d'échecs et mieux orienter les étudiants en fonction de leur niveau d'entrée (cf [1], [2], [3])

(2) Les classes préparatoires aux grandes écoles

Il s'agit d'une spécificité du système Français. Elles sont installées dans des lycées et l'organisation du travail ressemble beaucoup plus à celle du lycée qu'à celle de l'Université.

Elles accueillent environ 11 % des bacheliers principalement en Sciences (séries C - E et D) et Lettres (série A), après une sélection assez rigoureuse sur dossier. Actuellement plus de la moitié des bacheliers C , en général les meilleurs, rentrent en classe préparatoire. L'enseignement de mathématiques y est donné par un seul professeur agrégé, qui assure à la fois le cours, les travaux pratiques et les interrogations. Ce professeur ne fait en général pas de recherche. Pour plus de détails, on pourra consulter [4]. L'enseignement d'informatique, récemment introduit, est assuré par le professeur de mathématiques. Les programmes sont rigoureusement fixés au niveau national et correspondent à un concours d'entrée dans une Ecole d'Ingénieurs ou Ecole Normale Supérieure passé au bout d'une ou plus généralement deux années. Le nombre d'écoles est tel que plus de 90 % des candidats sont admis à l'une ou l'autre.

(3) Les Instituts Universitaires de Technologie

Placés administrativement dans les Universités, ils sont largement autonomes. Ils accueillent pour deux ans environ 13 % des bacheliers principalement scientifiques, industriels, ou de gestion. Les étudiants sont sélectionnés sur dossier. L'enseignement de Mathématiques y occupe un horaire réduit, nul dans certaines sections. Il est assuré soit par des mathématiciens, en général sous forme intégrée (cours - travaux dirigés), soit par les spécialistes des autres disciplines. L'horaire lourd (35 heures hebdomadaires) ne permet pas aux étudiants de travail personnel en dehors de stages et de projets où les mathématiques ont peu de place.

(4) Les Sections de Techniciens Supérieurs

        Les lycées techniques comportent des sections conduisant en
deux ans après le baccalauréat à un Brevet de Technicien Supérieur.
L'enseignement y est donné par des professeurs de lycée. Il y a de très
nombreuses sections et spécialités. Ces sections accueillent environ
20 % des bacheliers.

(b) Deuxième Cycle

(1) Université

        Environ 65 % des étudiants de premier cycle poursuivent leurs
études dans un deuxième cycle sanctionné par une maîtrise. Depuis quelques années se développent des Maîtrises d'Ingénierie Mathématique.
Les enseignements sont donnés par des enseignants-chercheurs titulaires
à plein temps. Certaines universités délivrent un diplome d'ingénieur.
En 1984-85 il y avait 47 557 étudiants inscrits dans une Maîtrise scientifique et 9 374 inscrits pour des études d'ingénieur.

(2) Grandes écoles

        Certaines sont intégrées à une Université (E.N.S.I. -
Compiègne) (12 064 inscrits en 1984-85), d'autres dépendent de l'Education Nationale mais pas d'une Université (13 425 inscrits), d'autres,
comme l'Ecole Polytechnique, d'un autre Ministère (9 191 inscrits),
d'autres enfin sont privées(10 055 inscrits).

        De même les écoles de commerce, privées pour la plus grande part,
accueillent 26 000 élèves à rapprocher des 26 000 étudiants inscrits en
2ème cycle de Sciences économiques à l'Université.

        L'enseignement dans ces écoles (cf [4] § 5) est en général donné
par un grand nombre d'enseignants à temps partiels issus de l'industrie.
Il y a très peu d'enseignants universitaires permanents (sauf en mathématiques à l'Ecole Polytechnique).

(c) Troisième Cycle

    La formation à la recherche et par la recherche est beaucoup plus
développée à l'université que dans les grandes écoles où peu d'élèves
achèvent leurs études par une thèse.

3. LA FORMATION PERMANENTE

La plupart des Universités possèdent un centre de formation perma-
nente qui organise à la demande des enseignements spécifiques. Ces enseignements sont rarement sanctionnés par un diplome universitaire.
A côté des universités, le Conservatoire National des Arts et Métiers
et ses 51 centres régionaux ont mis en place tout un système d'enseignements modulaires dispensés le soir et le samedi et ouverts à toute personne déjà engagée dans la vie professionnelle (environ 100 000 inscrits). Ils permettent de préparer des diplomes de différents niveaux
jusqu'au diplome d'Ingénieur ou au Doctorat. 400 diplomes d'ingénieurs
sont délivrés chaque année. Cinq départements réunissent près de 80 %
des diplomes de niveau DEUG ou Maîtrise : Mathématiques et Informatique
(26 %), Electronique - Electrotechnique et Automatisme (22 %), Economie
et Gestion (15 %), Chimie - Biologie (9 %), Travail et entreprise (7 %).
A Paris l'enseignement est donné par des enseignants-chercheurs à temps
complet. Dans les centres associés, l'enseignement est donné, soit par
des Universitaires, soit par des Ingénieurs.

4. LA RECHERCHE SUR L'ENSEIGNEMENT DES MATHEMATIQUES A L'UNIVERSITE

Les I.R.E.M. (Instituts de Recherche sur l'Enseignement des Mathématiques créés en 1968 se sont surtout efforcés d'étudier l'enseignement
des mathématiques dans le premier et le second degré. Toutefois il
existe une commission nationale inter IREM-Université qui s'est donnée
pour tâche de recenser et d'évaluer toutes les innovations concernant
l'enseignement des Mathématiques à l'Université (cf [5] et [2]).
L'université de Toulouse a organisé en 1982-83 un séminaire sur ce
sujet (cf [6]).

Une thèse d'état de didactique des mathématiques ([7]) porte sur
l'acquisition du concept de convergence des suites. Enfin la brochure
[8] concerne surtout les étudiants en formation continue (C.N.A.M.)

REFERENCES

[1] Journées de réflexion sur les premiers cycles rénovés
     I. Nice - 29-30 Mai 1986
     II. Nice - 21-22 Mai 1987

[2] Colloque inter IREM Rénovation des premiers cycles : le rôle des
     Mathématiques (12 et 13 Novembre 1987). Université de RENNES I
     Campus de Beaulieu 35042 Rennes Cedex

[3] Colloque "orientations et échecs" Paris-Dauphine - 22-23 Mai 1987
     Bilan des acquis scientifiques (D. Lacombe). Mathématiques p. 1 à
     90

[4] La France en Mai 1981 - volume 4 : l'enseignement et le développe-
     ment scientifique
     Contribution de Laurent Schwartz
     3. L'enseignement supérieur à l'Université (p. 241-303)
     5. Les grandes écoles   (p. 372-422)
     La Documentation Française. Paris 1981

[5] Enquête : points d'innovation dans l'enseignement des mathématiques
     à l'Université. Groupe Inter IREM UNIVERSITE. Irem de Rennes 1984

[6] Enseignement des Math après le Bac. Irem de Toulouse 1983

[7] L'acquisition de la notion de convergence des suites numériques
     dans l'enseignement supérieur. A. Robert
     Thèse d'Etat, Université de Paris VII, Juin 1982

[8] Obstacles et déblocages en mathématiques
     M. Bruston et C. Rouxel - Brochure APMEP n° 47 - 1982

MATHEMATICS SERVICE COURSES:
A CANADIAN PERSPECTIVE

B.R. Hodgson.
Université Laval, Québec, Canada

E.R. Muller
Brock University, Ontario, Canada

ABSTRACT

This paper is concerned with the general situation of mathematics service teaching in Canadian universities. It discusses university admission criteria and the age distribution of students within those institutions. It analyses the rôle of mathematics service courses and concludes that this rôle is surprisingly widespread and diverse.

1. INTRODUCTION

In response to the call by ICMI [1], we describe the general situation of Canadian universities with respect to admission and age distribution of students. Within this context we analyze the rôle of service courses by making use of data on Canadian mathematics departments which have been accumulated by the American Mathematical Society and by surveying mathematics departments in a number of different Canadian universities. We also include a section which reflects on some aspects of service teaching, an attempt to synthesize the issues raised in our other paper on "The mathematics service courses environment" [2].

2. ACCESS TO UNIVERSITY EDUCATION

Education is a provincial responsibility in Canada. This causes differences in syllabuses, graduation requirements and years of schooling from one province to the other. For example, most of the ten Canadian provinces have twelve years of schooling required for university admission; however, there are in Québec eleven years of primary and secondary schooling followed by another two years of college education in a CEGEP, while in Ontario the equivalent of thirteen years of schooling is required for university admission. It should be noted that most university undergraduate honours programs, except some professional ones like medicine, require four years of study, and three in the case of Québec (to compensate for the thirteenth year of preuniversity education). Students can thus typically receive an undergraduate degree at about age 22, after sixteen years of study.

The latest data available on education in Canada (for the year 1984-85; see [3] -- this report also presents a few condensed facts about the general structure of education in Canada) indicates that the total enrolment in university undergraduate programs was then 647 900, of which 404 200 were enrolled full-time. (The total population of Canada at that time was approximately 25 millions.) Postsecondary education can also be obtained in "community colleges" which offer career programs of one to four years (applied arts and technology, agriculture, etc.) as well as one- to two-year academic programs preparing a student to proceed to university (for example Québec's CEGEPs). The total 1984-85 enrolment in community colleges was 321 600, including 95 800 in university transfer programs. Total full-time postsecondary enrolment have increased steadily in the seventies and early eighties with a growth of 40% over a ten-year span.

The age distribution shows that 12,8% of 18-21-year-olds were enrolled full-time in university undergraduate programs (the proportion being 13,5% for female students) while another 1,1% was enrolled part-time. The proportion of the 19-year-old population enrolled full-time in postsecondary education (university and community colleges) was 28%. The following age distribution of the 243 700 part-time students in university undergraduate programs (38% of the total enrolment) demonstrate the importance of adults students and continuing education:

19 years and less	2%
20-24 years	23%
25-29 years	22%
30-39 years	33%
40-49 years	14%
50 years and over	5%

The above data demonstrates that university programs attract a large percentage of students who have been out of school for a number of years. The work experience and academic background of these students tend to vary very much. Furthermore students entering directly from secondary school (or Québec's CEGEPs) very often have substantial choice as to which courses to elect during their final year before university. However, admission to particular university programs generally impose restrictions on those choices. This has traditionally been the case for students aiming at a scientific discipline, the admission profile usually including mathematics (mainly calculus). More recently, however, some other programs have been requiring a special mathematics preparation for admission. For instance, the Ministry of Education in Ontario has just released a new Curriculum Guideline for Mathematics in the Intermediate and Senior High School Divisions [4] which take into account the increased university mathematics requirements for social sciences and related majors. In their final year of schooling, students will thus have the opportunity to select one or more of the following mathematics courses:

I   algebra and geometry
II   calculus
III   finite mathematics.

While options I and II have traditionally been seen as the main preparation for university mathematics, the finite mathematics course is now offered aiming at the preparation of students for the mathematics service courses in social sciences, etc. The demand for such a finite mathematics course will depend largely on the extent to which Ontario universities will require this course and whether or not students will perceive it as limiting their options for entry into university programs.

While entrance to university requires the student to choose the right options for his elected major, there is usually no special admission procedures (for example entrance examination) in Canada. It is true though that a few programs such as medicine or dentistry are highly selective (sometimes because of physical constraints like the number of dentist chairs or lab places available at the university) but, in general, admission to Canadian universities can be described as open, sometimes limited by indicators based on the final year before entering university.

3. MATHEMATICS SERVICE COURSES: SOME DATA

Undergraduate programs may vary substantially from university to university even for the same discipline. Within this multiplicity of undergraduate patterns one finds many implicity and explicity defined mathematics service courses. Institutions are independent and have developed differently, depending on the various challenges and pressures within the university, the composition of mathematics departments, faculty interests, etc. Therefore one finds service bonds between mathematics and different departments in the various institutions.

In the universities it appears that two factors predominate in influencing what service courses are explicity offered. These are
   (i)   the responsibilities of the mathematics department.
             Do they include one or more of statistics, computer science, operations research, actuarial science, etc.?
   (ii) the size of the institution.
             Independent units with their own mathematics courses are more prevalent at large institutions.

Sometimes in one institution a course will be considered as a regular mathematics course for students of the department while in another institution a very similar course will be classified as a service course -- for example a course in linear algebra offered mainly to computer scientists will be considered a service course if mathematics and computer science are two separate departments while it would not be in an institution where computer science is part of the mathematics department.

In order to get a better understanding of the Canadian situation with respect to mathematics teaching, we have studied the American Mathematical Society Annual Survey data. This data demonstrates some general trends. In Canada there are close to 100 universities; thirty-five of them offer PhD's in mathematics and of the remainder about half offer MSc's in mathematics. The majority of the PhD institutions were established before 1960 while the other universities were founded mainly in the early 1960's. Separate data for the institutions which do not grant PhD's in mathematics is available from 1977 to 1982 and is plotted in Graph 2. Graph 1 shows data for all institutions from 1973 to 1976 and only PhD granting institutions thereafter.

From Graph 2 we see that in the non-PhD granting institutions first year calculus accounts for 30% of total registrations in mathematics, precalculus courses for about 10%, statistics 18%. During this period computer science course enrolments dropped from 28% to 14% (as separate departments of computer science were formed in many institutions), and other undergraduate courses increased from 15% to 28% of total enrolment. In such institutions it appears that nearly all statistics enrolment is service course enrolment. In most provinces the

Graph 1: *All Institutions 1973-76; PhD Granting Institutions 1977-85.*

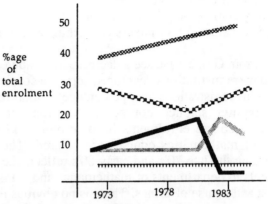

No Operations Research courses reported

Graph 2: *Non PhD Granting Institutions (separate category).*
(Data available for 1977 to 1982 only)

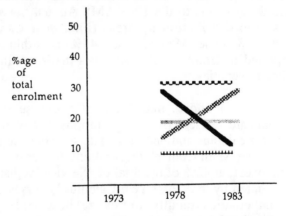

Some Operations Research courses reported, approximately 1% of total enrolment.

⊔⊔⊔⊔⊔⊔⊔⊔⊔⊔⊔⊔	**Below level of calculus**
∿∿∿∿∿∿∿∿	**First year calculus**
▬▬▬▬▬	**Computer Science**
∞∞∞∞∞∞∞∞∞∞	**Other undergraduate mathematics**
▓▓▓▓▓▓▓▓	**Statistics**

Data taken from AMS Annual Surveys 1973-1984 kindly supplied by Professor Donald Rung, The Pennsylvania State University. Note that not all Canadian Universities responded to the surveys, the data is aggregate and has been smoothed up to a maximum of ±4%.

courses below calculus level are taken before the student selects a major; in Ontario these courses are all service courses. Because of the particular structure of the schooling system in the province of Québec, first year calculus, being taught in the CEGEP, is considered pre-university teaching and so is not included in the data.

From Graph 1 we see a decrease in computer science courses after 1978 as more and more institutions developed separate computer science departments. First year calculus registrations as a percentage of total registrations are at their lowest when computer science course registrations are at their maximum. Other mathematics course enrolments become more and more significant as a greater diversity of mathematics courses are offered. The increase in statistics course enrolments in the late 70's and early 80's reflects the change in many mathematics departments to departments of mathematics and statistics not only in name but also in interest and course offerings. There is no obvious reason for the small decrease in statistics enrolments after 1983. We are struck by the importance and stability of first year calculus enrolments over the 10 year period.

To provide more details on the service component of course registrations, we surveyed some fifteen university mathematics departments throughout Canada (cf. acknowledgements). The survey consisted of a breakdown of the data submitted by these departments to the 1985 AMS Annual Survey. Table 1 summarizes our findings. The distribution by areas (First year Calculus: 37%; Undergraduate Statistics: 15%; Other Math Courses: 48%) is within the 4% variation of the AMS data reported in Graph 1. This suggests that our sample is representative of the Canadian universities.

Table 1 shows that 4 out of 5 course registrations in mathematics departments are service course registrations. We were surprised to find that when the first year calculus and undergraduate statistics are removed from consideration, 70% of "Other Math Course" registrations are service course registrations. In fact, this component, at 42% of the total of "Service" registrations, is comparable to the 43% accounted by the "First Year Calculus" courses. It may come as a surprise to many mathematics faculty to realize how widespread and diverse service mathematics has become in Canadian universities.

Table 1:   *Course Registrations - Departments of Mathematics and Statistics*

Table gives percentages averaged over the universities surveyed (over 13 000 math courses registrations).

	Math Majors	Service (Other Majors)	Total
First year Calculus	5 *10*	95 *43*	100 *37*
Undergraduate Statistics	19 *15*	81 *15*	100 *15*
Other Math	30 *75*	70 *42*	100 *48*
Total	19 *100*	81 *100*	100 *100*

4. SOME ASPECTS OF SERVICE TEACHING

It is generally recognized that in a mathematics department its own program benefits from the service teaching component. Because of the large enrolment and high student-faculty ratio in service courses, it is then easier to have faculty involved in the smaller (sometimes very small!) enrolment upper year mathematics courses, in graduate teaching and also in research. For example at Université Laval (Québec), service teaching account for more than 85% of all courses *registrations* within the mathematics department but for only half of the courses *sections*. (Moreover more than a third of service courses sections are actually taught by part-time instructors.) It is true of course that courses enrolment is not the only measure by which human and financial resources are distributed among the different departments in a university. But still these considerations stress heavily the fact that any mathematics department which neglects its service teaching is likely to find itself with diminished resources for teaching to its own undergraduate and graduate students.

However such benefits come at a price! One must not underestimate the costs to the department and, more specifically, to the faculty. These costs involve such things as loss of autonomy, external evaluation, less interesting but more administratively demanding work, work for which little university credit is given, etc. It is therefore crucial that mathematics departments develop their own policy regarding service courses and that these be communicated to the appropriate administration to ensure that due credit is given for the important work they perform in this area.

The recent development of teaching oriented software in the areas of calculus, linear algebra, statistics and operations research will have an important impact on the mathematics service courses. This impact has already been felt in service statistics courses. Results of our Canadian survey demonstrate that approximately 40% of all students registered in statistics courses offered by mathematics departments are required to use computer facilities (this average may not be very representative as the range of computer usage is reported from 0% to 60%). However, only one university reported required computer usage in other mathematics courses above 5%. No university reported required computer usage in its first year calculus courses.

It is our opinion that teaching oriented software should be seriously considered for all service courses. The introduction of such software requires careful planning and can eventually bring substantial changes in the course structure. It is a time consuming job and those faculty who have experience in this area should communicate their findings to the mathematics community.

A more detailed discussion of the implications of mathematics service course teaching in Canada is presented in [2].

5. CONCLUSION

The analysis of existing surveys and our own limited survey indicate that mathematics departments in Canada service a widespread and diverse clientele. It also demonstrates a heavy dependence of mathematics departments on service courses enrolments. This dependence has surprised the majority of faculty we have spoken to.

Limited experimentation with teaching oriented software appears to indicate that it can play an important rôle in service teaching. More experimentation is needed and the results should be communicated to the many faculty involved in service courses.

ACKNOWLEDGEMENTS

We are grateful to the following professors for the data they supplied on their departmental service course enrolments:

R.M. Barron (*University of Windsor*), P. Berthiaume (*Université de Montréal*). P. Browne (*University of Calgary*), J.M. Geramita (*Queen's University*), J. Macki (*University of Alberta*), M. Malik (*Concordia University*), J. Poland (*Carleton University*) and J. Wick Pelletier (*York University*).

REFERENCES

[1] Howson, A.G., J.-P. Kahane, P.J. Kelly, P. Lauginie, T. Nemetz, F.H. Simons, C.A. Taylor and E. de Turckheim, "Mathematics as a service subject." *Enseign. Math.* 32 (1986) 159-172.

[2] Muller, E.R. and B.R. Hodgson, "The mathematics service courses environment." In: Clements, R.R. and A.G. Howson (eds.) *Selected Papers on the Teaching of Mathematics as a Service Subject*. (ICMI Symposium, Udine, 1987) Springer-Verlag.

[3] *Education in Canada: A Statistical Review for 1984-85*. Statistics Canada, 1986. (Catalogue 81-229 Annual)

[4] *Curriculum Guidelines for Mathematics in the Intermediate and Senior High School Divisions*. Government of Ontario, Ministry of Education, Toronto, Ontario, 1985.

MATHEMATICS AS A SERVICE SUBJECT —
THE AFRICAN EXPERIENCE

A.O. Kuku
University of Ibadan, Ibadan, Nigeria

ABSTRACT

The teaching of mathematics as a service subject raises similar problems in almost all countries. Such issues as who should design the syllabus for the service courses and who should execute the instruction of the courses are universal. The particular problems faced by African universities, especially shortages of staff and other resources, add further facets to the universal issues. This paper discusses the African experience in the teaching of service mathematics.

1. INTRODUCTION

We are aware that mathematics and mathematical
sciences these days encompass not only the core components
like algebra, analysis, topology and geometry but also a
wide range of subjects like computer science, operations
research, statistics, econometrics, biomathematics,
mathematical physics, chemistry, linguistics or even
mathematical history otherwise known as cliometrics. Whilst
many of the subjects so far mentioned have mathematical
characters of their own, and are also serviced by other
areas of mathematics, examples abound of various areas of
science and technology - agriculture, engineering, social
sciences - where mathematical sciences are rendering
invaluable and sometimes unpredictable services.

There are numerous examples of situations where a
mathematical science has played a service role not only to
another mathematical science but to other sciences, whilst
some mathematical sciences have led to further developments
in the core areas of mathematics. Instances of such
unpredictable cross-fertilization of ideas include
[i]   the recent use of elementary results of projective
        geometry by N Karmala of Bell Laboratories, USA to
        discover new techniques for solving linear programming
        problems,
[ii]   the emergence of differential geometry as a natural
        mathematical framework for the gauge theories of
        physics - a situation that has also led to the
        discovery of an exotic Euclidean 4-space whose study
        will no doubt point to new insights into the topology
        of four dimensional spaces,
[iii]   the use of data analysis in computer science to study
        epidemiological problems and the use of computers in
        diagnosis which provide automatic blood and urine
        analysis as well as tomography of internal organs,
[iv]   the use of the simplex method in linear programming to
        facilitate the planning of industrial production,
        inventory control and hence the most efficient
        allocation of resources,
[v]   the use of number theory in national security matters
        where prime numbers form a basis for new schemes for
        constructing or analysing cyphers, and
[vi]   the use of Fourier analysis to study electro-magnetic
        waves and their harmonic components (eg X-rays), and
        the construction of many electrical and electronic
        devices such as nuclear magnetic resonance
        spectrometers, X-ray crystalographic spectrometers,
        etcetera. Moreover, Fourier analysis has provided a
        basis for understanding quantum theory and hence
        modern chemistry and physics as well as proved useful

in time-series analysis used in oil exploration for
interpreting seismic waves transmitted through rocks
suspected of bearing petroleum.

The multitudinous uses of mathematical sciences in
other areas of science and technology make mathematics a
very important service subject and the aim of this paper is
to assess the situation of the teaching of mathematics as a
service subject on the African continent. In trying to
assess this situation, the following questions readily came
to mind.

[1] How do the structures of the teaching units for
    mathematical sciences in African universities affect the
    teaching of service courses?
[2] Should service courses be designed by mathematics or
    user departments?
[3] Who should teach mathematics service courses -
    mathematicians or users of mathematics?
[4] Should the teaching of service courses be done in
    mathematics departments or user departments?
[5] What type of service courses are being taught and what
    improvements are desirable?

We shall try to discuss and provide some answers to the
above questions.

2. THE EDUCATIONAL SYSTEM IN AFRICA - NIGERIA AS A CASE STUDY

The schooling systems in Africa vary slightly from one
linguistic region to another. However, it does appear as if
many African countries are moving towards or are already
adopting a 6-3-3-4 system, ie six years of primary, three
years of junior secondary, three years of senior secondary
and four years of tertiary education of varying periods
where a first university degree is not less than four years
duration.

In Nigeria for instance, the latter system, 6-3-3-4, is
currently just being adopted in preference to the former 6-
5-2-3 system comprising six years of primary, five years of
secondary, two years of pre-university (or advanced level or
sixth form) training and three years of university education
to get a first degree. The major differences between the
two schemes are

[i]   the 6-3-3-4 system is intended to keep university
       education within the confines of the university since
       the advanced level or sixth form programme were
       usually part of the university training and most of
       the work at this level is not so relevant to most
       career options open to secondary school leavers, and

[ii]   whereas the five years of secondary education in the
        6-5-2-3 system limited the children to purely
        academic training, the 6-3-3-4 system is intended to
        expose students in the junior secondary to a wide
        variety of usable options with a view to making them
        identify areas - academic or technical - which they
        could pursue at the senior secondary level; indeed
        this system is intended to diversify the secondary
        school curriculum to cater for the differences of
        talents, opportunities and roles possessed by or open
        to students after their secondary school careers.
   The problem facing the implementation of this new
system is typical of what happens in many African countries.
Shortly before the introduction of the 6-3-3-4 system, the
Federal Government had, in 1976/77, introduced universal
[free] primary education, which increased the primary school
enrolment in the country from 6,165,547 in 1975/76 to
14,311,608 in 1981/82 (by which time the programme was in
its sixth year) - a 132.12% increase. This, of course, has
also meant a phenomenal increase in the cost of education at
this level - facilities, buildings, training of teachers,
etcetera - which is now estimated at an average cost of ₦4
billion [one billion per year].
   Most of the students leaving primary school in 1982 had
to find places in the new junior secondary schools built at
enormous costs to the Government. By 1985 many of the
students in the junior secondary schools were promoted to
senior secondary schools and the first products of the
senior secondary schools under this system will finish their
secondary education in 1988.
   The main problem with the secondary school aspect of
the programme has been the purchase of workshop equipment
for courses in Introductory Technology to facilitate the
technical aspects of student training at this level. Since
the introduction of the new system, new mathematics books
have had to be written for the junior and senior secondary
schools and the response to this challenge, from individuals
and associations, has been tremendous. Indeed the
Mathematics Association of Nigeria has succeeded in
organising its members to write all the books at this level
in the name of the Association.
   Most of the students terminate their education at the
senior secondary level when they complete their secondary
school leaving certificate. However, those who wish may
take the University Entrance Examination organised by the
Joint Admission Matriculation Board [JAMB] and successful
candidates in this examination spend the next four years
studying for a university degree. Others may choose to go

into colleges of technology or polytechnics or colleges of
education.

3. TYPES OF MATHEMATICS TEACHING UNITS IN AFRICAN UNIVERSITIES

Before discussing the problems associated with
mathematics as a service subject in Africa, it seems
appropriate to give some insight into the teaching units
mostly responsible for the service courses.

Basically, there are three major categories of
universities where programmes in mathematical sciences are
offered and every university invariably offers some service
courses. Category A consists of universities whose
mathematics departments have statistics and computer science
as components and have students majoring in mathematics,
possibly with specialization in statistics or computer
science as well as offering service courses in mathematics.
Such universities include Ahmadu Bello University, Zaria,
Nigeria, Universities of Nairobi, Abidjan, Younde, Dakar and
Lesotho. Category B are universities whose mathematics
departments, like category A, have statistics and computer
science as their components but offer only service courses
without specialist degrees in mathematics, except probably
mathematics education. Examples of such universities
include Universities of Zambia, Malawi, Mozambique and
Swaziland. Category C consists of universities with
separate departments of mathematics, statistics and computer
science, with specialist programmes in each of mathematics,
statistics and computer science as well as service courses
in mathematics. Such universities include Universities of
Ibadan and Nsukka in Nigeria.

One common feature of universities in Africa is that
most of them are inadequately staffed as far as mathematical
teaching is concerned. Indeed, the universities in
categories A and C have to struggle to cope with the heavy
demands on their staff for specialist and service programmes
as a result of which the service programmes suffer quite a
lot in terms of quality of instruction, inevitability of
large classes, inadequate tutorials, etcetera. To compound
matters, many of the universities in categories A and C
offer postgraduate programmes which their meagre staff to
giving graduate courses thus stretching their human
resources beyond reasonable limits.

The systems of instruction in African universities vary
from country to country. However, many of the African
universities (eg Ibadan, Ife, Nsukka, Lagos [all in
Nigeria], Addis-Ababa, Nairobi, and others) have adopted the
course unit system, with mathematics courses broken up into

small modules lasting one semester, and where the courses
have to be examined at the end of the semester in which the
courses are taught, with obligations for the staff to give
tests for the students almost every month. For this system
to work efficiently, there should be a large number of
courses to cater for various needs of specialists and users
with equivalent requirement for a large number of staff as
usually happens in mathematics departments in the USA [for
instance the Mathematics Department at the University of
Illinois at Urbana has over 120 members on its staff]. The
limitation of staff in many of the African universities
operating the course system invariably affect negatively the
mathematics service courses offered by these universities
since classes for service courses are invariably too large.
       Another feature of a well operated course unit system
is the availability of a large number of Graduate Assistants
for tutorial purposes - a system that presupposes the
existence of a viable postgraduate programme with a large
number of postgraduate students from whom Graduate
Assistants could be appointed. Sadly enough, many of the
universities offering mathematics service programmes have no
viable graduate programmes and those which do usually do not
have enough students to appoint as Graduate Assistants and
even where they have enough postgraduate students, financial
limitations usually prevent the appointment of such Graduate
Assistants.

4. WHO SHOULD DESIGN SERVICE COURSES - MATHEMATICIANS OR USERS?

       The question of who designs mathematics service courses
has generated debates and controversies in several African
universities. To the best of my knowledge, most service
courses in African universities are designed by the
mathematics teaching units in the universities with or
without consultation with the departments to which the
services are being rendered. Many a time, it is a matter of
the user departments looking at what mathematics service
courses are available in the mathematics departments and
fitting into them.
       A major defence for this practice is the view that many
user departments have no clear ideas about the mathematics
they need because of the mathematical limitations of their
staff members. It is also usually argued that a good user
programme is one which, apart from providing the user of
mathematics with his current needs, also gives him a bit of
extra knowledge of mathematics for his future needs.
       Even though many lecturers in user faculties like
agriculture don't have much by way of mathematics

background, the same cannot be said of lecturers in engineering or economics departments since many of them are quite knowledgeable as far as mathematics is concerned.

However, there is one other reason that makes it difficult to implement the results of full consultation with user departments, namely that the mathematics department feels it necessary to design courses for groups of user departments rather than individual user departments because of their staff size limitations which makes it necessary to limit the number of courses to a manageable number.

Granted that most African universities cannot afford to design special service courses for each user department, it would seem more appropriate, in order to improve the quality of service courses, that there should be more consultations between mathematics departments and the user departments in such a way as to ensure that, while courses are designed for groups of user departments, each user department should benefit and classes for the service courses should not be too large.

5. WHO SHOULD TEACH SERVICE COURSES - MATHEMATICIANS OR USERS?

One other problem which has generated debate in African universities is whether Mathematics for Chemists should be taught in the Chemistry Department by a chemist, Mathematics for Economics in the Economics Department by an economist, Mathematics for Engineering in the Engineering Department by an engineer. The complaint has always been that service courses taught by mathematicians whether based in mathematics or user departments are usually too abstract to generate the interest of and be of use to the users.

However, in spite of protests by user departments, most of the mathematics service courses being taught in African universities are being taught by mathematicians based in either mathematics departments or user departments. There are various reasons for this. Firstly, as already pointed out above, the service courses are designed most of the time by mathematicians who believe also that they are best qualified to teach the courses. Secondly, in universities like Ibadan operating the course system, there is usually a rule that courses have to be taught in the departments housing the disciplines in order to avoid duplication of courses and maximise the use of university staff across departments and faculties. Thirdly, most user departments usually do not have academic staff qualified to teach the mathematics their students need.

It would appear that the agitation by user departments for relevance in the teaching of service courses needs a lot

of attention by mathematicians who usually teach the service
courses. It is advisable that those who teach service
courses should endeavour to motivate what they teach through
examples and illustrations relevant to the user departments
while maintaining some measure of rigour in the exposition
of the material for the courses. The tendency in many
African universities in categories A and C, owing to
shortage of staff, is to entrust the teaching of service
courses to young and inexperienced lecturers, sometimes
Graduate Assistants, while the more experienced members of
staff teach the honours or specialist courses. The argument
usually advanced in favour of this approach is that the
service courses are easier to handle. Yet we know that it
sometimes takes an experienced lecturer to create relevant
examples and make the lecture interesting to students. That
is why at Ibadan, each senior member of the mathematics
department is supposed to teach at least one service course
per session.

6. TYPES OF MATHEMATICS BEING TAUGHT AS SERVICE COURSES

Most science-based faculties in African universities do
require a pass in elementary mathematics - arithmetic,
algebra, Euclidean geometry, trigonometry - at credit level
for English speaking countries [or equivalent level for
French-speaking countries] for university admission.
However, for those who have not done additional mathematics
- calculus, algebra, analytic geometry, mechanics, and some
statistics - in school, and for many of science-based
faculties like Agriculture, Social Sciences, it is usual to
design a global supplementary mathematics course for new
entrants into the universities to update their mathematics
knowledge in preparation for their university studies. The
contents of such a course include some calculus, analytic
geometry, mechanics, further algebra, trigonometry and
statistics.
Now, to specialise in various science-based subjects,
like physics, chemistry, economics, all types of
engineering, etcetera, requires more knowledge of
mathematics than elementary calculus, analytic geometry and
so on, hence the need for students to take service courses
in various areas of mathematics. For universities in
categories A and C a few of these courses could be the same
as those for majors in mathematics while the majority are
designed specifically for users of mathematics. Many a
time, two courses with the same title may be designed, one
for the majors in mathematics and the other for non-majors
[users of mathematics].

At Ibadan, for instance, various types of courses for majors in mathematics have their counterparts for non-majors eg Abstract and Linear Algebra, Differential Equations, Analysis, Numerical Analysis, Complex Variables and such courses are attended by users from a group of departments or faculties, eg Physics, Chemistry and Engineering. However there are a few of the service courses also taken by majors in mathematics [eg Mathematical Methods, Special Functions, Vectors and Tensors, Operations Research and Mathematical Modelling]. Moreover, there are courses designed for specific users eg Mathematics for Social Sciences. This pattern is common to many English speaking countries. However, in Ibadan, in spite of the effort to maximise the use of academic staff through harmonisation and non-repetition of courses, it has been found inevitable to have, for instance, operations research courses in the Mathematics, Economics, Statistics and Industrial Engineering Departments.

The main reasons why universities offering specialist programmes design two types of courses [possibly with the same title] - one for specialists and the other for users - are

[1] the style of teaching specialists is different from that of teaching users of mathematics. In the specialist classes, mathematical results are rigorously proved while the users' classes lay more emphasis on how to use the results,

[2] the specialist classes have relatively few students and are able to move faster whereas the classes for service courses are usually large containing students from diverse backgrounds and motivations and invariably move at a slow pace, and

[3] the actual contents of the courses for specialists are usually different from those of courses for users even when they appear under the same titles.

On the whole, the major problems facing the design of service courses on the African continent are created by the insufficiency of teaching staff. As a result the number of courses available has to be seriously limited, with the classes rather large and the resultant problems of ineffective teaching, lack of tutorials, and consequent poor performance of students. There is no doubt that various service courses being given at the moment could be improved upon and new ones introduced, but most of the universities in Africa in categories A and C could hardly do this without negative effect on their honours undergraduate and postgraduate programmes, which are also esential for the various departments concerned.

SUMMARY OF THE HUNGARIAN EDUCATIONAL SYSTEM WITH SPECIAL ATTENTION TO MATHEMATICS TEACHING

T. Nemetz
Hungarian Academy of Sciences, Budapest, Hungary

ABSTRACT

In this paper the Hungarian educational system is described with particular reference to mathematical education. Detailed attention is then given to the place of mathematical service courses within university studies and, finally, the service mathematics syllabus for engineering courses is described.

1. PRE-UNIVERSITY EDUCATION

It is compulsory, in Hungary, for children aged between 6 and 14 to attend the so-called general school (primary school), which consists of grades 1-8. After finishing the general school about 80% of pupils continue their studies in secondary schools. There are three types of secondary school in Hungary.

a) Secondary schools of a general kind (called "gymnasium"), with four grades.
b) Secondary schools which are specialised in certain professions (eg textile industry, fine arts, etcetera), with four grades.
c) Training colleges with three grades.

Schools of type a) give a general education which enables the students to continue their studies at higher levels. Those of type b) also provide a general education but, at the same time, an expressed goal of the school is to enable the students to work towards the particular vocation chosen. They, too, might enter universities. The goal of the schools of type c) is to train skilled workers whilst enlarging their general knowledge. A currently changing aspect of the Education Act is that there has been a very detailed syllabus prepared by the Ministry of Education both for the general and secondary schools, prescribing for the teachers what to do almost lesson by lesson. The teachers must adapt themselves to this syllabus. (During the 1950s the number of students was increasing rapidly and teacher training could not keep pace with rising demands, therefore the introduction of such a syllabus was an imperative necessity at that time.)

Obviously such a system does not function in the case of more talented students. Therefore there are classes organised for them which are specialised in given subjects. Teachers teaching in these mathematics classes are well versed in higher mathematics and experts in methodological questions. Accordingly they enjoy much more freedom in their work.

The education of these more talented students is aided by nation-wide contests. These contests are well respected; for instance, for the first ten students in the National Contest, tertiary institutions (universities and colleges) waive the entrance examinations. In addition to that, in mathematics, there is a monthly periodical devoted completely to school children with an all-year-round column given over to problem solving contests.

2. TEACHER TRAINING

In Hungary, teachers, after successful graduation from secondary school, are trained at three different levels.
- Teachers for grades 1-4 in general (primary) schools are trained in teacher training colleges for three years. There the present program deals mainly with methodological questions, and hardly goes beyond the secondary school subject matter in mathematics.
- Teachers for grades 5-8 in general schools are trained in "Institutes of Higher Pedagogics" where the course lasts four years. There the students specialise in two or three subjects. Those who choose to specialise in mathematics become acquainted with the rudiments of higher mathematics.
-Teachers for secondary schools are trained in universities. They take two subjects. During their five years of education those students who specialise in mathematics are supposed to get acquainted with higher mathematics. As a criticism it could be said that they learn quite a bit of theoretical knowledge, but they know nothing about the applications.

3. MODUS OPERANDI OF THE HUNGARIAN UNIVERSITIES

When applying for admission to a university, applicants must fully specify their intended profession as one of those listed by the Ministry of Education. For all branches of study there is a "numerus clausus", that is the number of students is limited in all branches, all universities and all years. In order to enrol, students have to pass an examination. Once enroled it is nearly impossible to modify the branch of study chosen. Each of the five years of education comprise two semesters. Each semester has an instruction period, during which the courses are given, and an examination period, when the students have to pass compulsory examinations prescribed for their branch of study. There are no substitutes for these examinations. Failing an examination at the first sitting, one can attempt it a second time within the same period. Given special permission, which is generally granted, one more attempt is possible. Enrolment for the next semester is conditional on passing all compulsory examinations of the previous semester. Successful examinations are rated by one of four marks, which are of importance only for financial funds. The examinations generally consist of both written and oral parts. Usually the written examination in mathematics means solving problems.

An important feature of tertiary education in Hungary
is that "undergraduate" and "graduate" courses are not
distinguished, nothing corresponding to the title "graduate
student" exists. University studies terminate by obtaining
a "Diploma" having successfully defended a thesis prepared
by the candidate.

4. SERVICE MATHEMATICS

Mathematics service courses are offered during the
first four semesters with few exceptions. Here the word
"offered" is used in a very broad sense. In most cases the
course which is offered is compulsory for the students. As
a general rule, such courses are planned and offered by the
mathematics department.

The syllabus is (usually) compiled by the department(s)
of mathematics, taking into account the demands of the given
discipline (in chronological matters as well as in content)
and consulting with lecturers from the department concerned.
The lecturers for these courses are drawn from the
mathematics department and the tutors are from the special
fields. Exceptions to this general rule are biometry and
probability for physics.

A different policy is practised by the K. Marx-
University of Economics. There every student learns
mathematics at least during the first three semesters.
There are, however, disciplines which are strong in
mathematics, in which there are compulsory mathematics
courses during all semesters. The mathematical content of
these courses is fixed by a special committee appointed by
the university. Besides mathematicians, other departments
are represented here. Practising economists from outside
the university are also invited. Considerable parts of the
mathematics courses are given by economist-members of other
departments. A descriptive statistics course is offered
essentially only at this university, and it is almost taboo
for mathematicians.

5. MATHEMATICS FOR ENGINEERING STUDENTS

The compulsory mathematical subject matter for
engineering students is universal. During the first three
semesters analysis and algebra are taught. In the fourth
semester probability calculus appears - this being, though,
just an introductory course. The analysis courses include
the differential and integral calculus of functions of one
and several variables, ordinary differential equations,
complex functions (Cauchy forms, Laplace and Fourier

transforms), vector analysis (up to Stokes theorem) and
partial differential equations. Algebra comprises vector
algebra, matrix calculus and determinants.

The compulsory mathematics component in geo-sciences,
physics, chemistry and meteorology includes essentially more
theoretical _and_ application oriented subjects, since it is
offered at the Eötvös-University of Sciences, Budapest.

The subject of informatics will not be discussed in
this paper. Roughly speaking, it can be characterised
currently as being in a state of "catching up", at the pre-
university level as well as at the university level.

Finally it should be noted that the aim of teaching
mathematics as a service subject is seen differently by
different groups. Mathematicians insist that the goal is
training of the mind _and_ providing applicable knowledge at
the same time. Lecturers from other fields feel that the
aim of teaching mathematics is to familiarise the students
with the necessary mathematical theory and technique by the
time they need it in their main courses. The emphasis is
put on technique and students are even encouraged to prefer
formal thinking.

THE SERVICE TEACHING OF MATHEMATICS IN BRITAIN

J.C. Newby
Brunel University, Uxbridge, England

ABSTRACT

The educational system in Britain and the means of access to tertiary
(higher) education are described and figures given for the national
participation rates at various levels. The requirements of mathematics
service teaching to engineers, scientists, social scientists and the
arts are discussed together with the teaching methods employed and the
problems encountered. Finally, the increasing shortfall of potential
engineers, mathematicians and scientists, and of the teachers to teach
them, is highlighted.

THE BRITISH EDUCATIONAL SYSTEM

In Britain, all children must, by law, attend full time education from the ages of 5 to 16. The initial stage, between the ages of 5 and 11, takes place in primary schools. The vast majority of these schools are run and funded by Local Education Authorities (LEAs) on a regional basis with money coming to the LEAs from central government and from local rates. A small number of primary schools are independent and charge fees. These independent schools may have been set up for religious or other such reasons. All schools are subject to inspection by Her Majesty's Inspectorate of schools (HMI) who report to the Department of Education and Science (DES).

At about 11 years of age all children pass into secondary schools, the majority of which are again run by LEAs. As with primary schools there are some independent secondary schools which are self supporting and charge fees. The transfer between primary and secondary state schools is controlled by the LEAs which have considerable discretion. Different LEAs are at liberty to use different criteria. In the majority of regions all ability ranges transfer to a local comprehensive (secondary) school. In some regions there is a selection process which, for the more academically able children, leads to grammar (secondary) schools with the remainder going to the local comprehensive schools. Together with the independent schools this provides a range of educational possibilities at secondary level which varies considerably from one part of the country to another. As with primary schools the secondary schools are subject to inspection by HMIs.

As stated above, children in Britain are compulsorily educated up to the age of 16. At this time students, in consultation with parents and teachers, can decide to leave school and seek employment or to remain in full time education. During the run up to this period students will have been prepared for, and taken, a series of examinations to obtain some formal paper qualifications. This part of the British educational system is currently in a state of change. Previously the less academic students would have been encouraged to aim for passes in the Certificate of Secondary Education (CSE) examinations with the more academic taking General Certificate of Education Ordinary Level examinations (GCE "O" Level). There is some overlap with a CSE grade 1 being equivalent to an "O" Level pass. A new General Certificate of Secondary Education (GCSE) is currently being introduced to replace both CSE and GCE "O" Levels. The CSE and GCE examinations are run by regional Examination Boards who construct syllabuses and organise the examinations. They have considerable autonomy in this area and are free to introduce innovative changes subject to approval by the Secondary Examinations Council, a government run overseeing body. In this respect there is a healthy diversity at this level which can seem confusing to an outsider. The new GCSE system is again organised on a regional basis, the regions tending to be an amalgamation of various CSE/GCE regions resulting in fewer, but larger, regions. They have a similar autonomy to the CSE/GCE boards.

ENTRANCE TO TERTIARY EDUCATION

Students who stay on at school after the age of 16 might be seeking to improve their CSE or GCE "O" Level results but the majority will be studying for entrance to the tertiary educational system of Training Colleges, Colleges of Further Education, Technical Colleges, Polytechnics and Universities; the latter two being capable of granting degrees. Entrance to polytechnic and university education is normally achieved by obtaining appropriate grades in GCE Advanced ("A") Level examinations at the age of 18. Each polytechnic and university department can set its own admissions criteria. Typical requirements might be three "A" Level passes with grades of no less than B, C and D, the B being required in mathematics for entrance to a Mathematics or Engineering degree course. With the diversity among the GCE Boards and the admissions autonomy of the polytechnics and universities in Britain, the possibility arises of a mismatch between a student's academic background in mathematics and the course assumptions about this background. This potential problem is overcome by the existance of an agreed "common core" of mathematical topics which are included in all "A" Level mathematics syllabuses. In Scotland the situation is slightly different in that students there take the Scottish Higher Certificate of Education at the age of 17 and enter Scottish universities a year earlier than in England and Wales. Because of this the Scottish university degree courses are normally of four years duration, while those in England and Wales are normally of three years duration. Even here there are some exceptions. A few universities offer courses which include, as an essential part of their degrees, one or more periods of work in industry, commerce or government departments. At the author's own institution, Brunel University, Uxbridge, there are three such periods, each of six months, which form an integral part of a four year degree course.

PARTICIPATION IN TERTIARY EDUCATION

Participation rates in tertiary education can be notoriously difficult to compare between countries due to different national interpretations of terms. In the International Standard Classification of Education (ISCED) handbook tertiary (higher) education is defined to be further specialised study normally undertaken after successful completion of a good basic education lasting for at least eleven years. Within tertiary education three sub-divisions are recognised, namely levels 5, 6 and 7. Level 5 courses are below degree standard and tend to be vocational. In Britain these include Higher National Diploma and Certificate (HND/HNC) courses and nursing qualifications. First degree courses come under level 6 with postgraduate programmes at level 7. Even within these definitions there is scope for national interpretations. Using these, and Organisation of Economic Cooperation and Development (OECD) standards, the DES in Britain has produced Statistical Bulletin 4/87 (1) in which comparisons in higher education

are made between France, Germany, Italy, Japan, the Netherlands, the
United Kingdom (UK) and the United States of America (USA).

In 1984, the year upon which the DES based its survey, 70% of new
entrants to higher education in the United Kingdom were in the 18-21 age
group, giving a participation rate of 31% for this group. There was a
total enrolment (total student count) of some 1007000 (35% women), of
which 35% were at level 5, 54% at level 6 and 11% at level 7. Of this
total enrolment 42% were in the university sector. Also in 1984 some
284000 students gained qualifications with 14% being in the sciences and
16% in engineering.

At any one time there are about a quarter of a million full time
undergraduates in UK universities with the majority of these requiring
at least some mathematics. In general those entrants to Mathematical
Studies, Engineering and Physics are required to have an "A" Level
qualification in mathematics, whereas a minimum of an "O" Level pass in
mathematics is required of entrants to the Biological Sciences,
Chemistry and Economics. Indeed an "O" Level pass in both mathematics
and English are general requirements for entrance to all university
courses. In broad terms, therefore, service courses of two types are
required. Firstly, for those entrants who have an "A" Level
qualification in mathematics, and secondly, for those with an "O" Level
qualification. The "O" Level qualification is essentially pre-calculus
whereas calculus and differential equations are part of the "common
core" at "A" Level.

SERVICE TEACHING TO ENGINEERS

The student engineer will normally have a good "A" Level grade in
mathematics. To become a professional engineer he or she will require a
degree and membership of the appropriate engineering institute
(Electrical, Mechanical, Civil). In order to qualify for institute
membership a student must demonstrate proficiency in a number of
specified subject areas. One way of demonstrating this proficiency is
to follow a university engineering course whose structure and syllabuses
have been examined and approved by a particular institute as providing
the requisite proficiency. In this way the engineering institutes can
have a considerable influence on engineering degree courses. As far as
the service mathematician is concerned this is reflected in the time
made available for mathematics and indirectly in the content of the
mathematics syllabuses. Although there are considerable variations from
one course to another, and from one university to another, a figure of
about 6 hours per week devoted to mathematics in the first year of the
course and 4 hours per week in subsequent years would provide a rough
average. The material taught is essentially mathematical methods. In
the first year such topics as vectors, matrices, linear algebra,
calculus, differential equations, Laplace transforms and elementary
probability and statistics would be covered. Students would have met
most of these topics at "A" Level but not to such a depth. Other topics

which might be included are numerical analysis, Fourier series and
Fourier transforms, and functions of more than one variable. All
students will use computers from very early on in the course with the
majority having had access to computers at school. Second year topics
are likely to include partial differential equations (including
numerical methods), complex analysis, vector calculus and statistics.
Specialist topics such as control theory and electromagnetic theory,
although frequently taught by mathematicians, tend to be well integrated
with the corresponding engineering applications.

Engineering departments are some of the largest departments within a
university and their first year mathematics courses can be common to
more than one department. As a consequence a lecturer can be faced with
very large classes indeed. It is, however, generally agreed that the
students are fairly well motivated, more so than some mathematicians,
and are therefore seldom a problem to teach.

SERVICE TEACHING TO SCIENTISTS

Science departments, such as Physics, which have "A" Level
mathematics as an entrance requirement are service taught mathematics in
a similar manner to engineers. For departments which do not have this
requirement, such as Biology, the teaching has to start at a
pre-calculus level. In the majority of cases the students will not have
done a course in mathematics for at least two years and may well have
lost what manipulative abilities they previously had. Such students
have to be reintroduced to mathematics rather gently if significant
progress is to be made. Manipulation of fractions and indices together
with an introduction to exponentials (biological growth) and logarithms
(pH values) can often provide a starting point. The service course will
generally progress through the binomial theorem, elementary probability
and statistics, matrices, graphs and equations of common functions,
differentiation and integration, ordinary differential equations,
partial derivatives and partial differential equations. The time
allocation is significantly less than that for engineers and might
typically be 2 or 3 hours per week for chemists. The teaching of
mathematics seldom goes beyond the second year and for some biologists
it might be for one year only.

The major problem encountered by the lecturer is one of lack of
motivation. Having not used much, if any, mathematics between "O" Level
at age 16 and "A" Level at age 18 for their primary subject, the
students cannot see why it is now required. In order to increase
motivation the mathematics lecturer must become reasonably familiar with
the subject area for which he or she is providing a mathematics service,
and introduce as much mathematics as possible by using applications in
this subject area.

As with engineers, these students are likely to have been introduced
to computers at school and will do a computing course early on in their
studies at university.

SERVICE TEACHING TO SOCIAL SCIENTISTS

Students of Economics and Psychology normally have a service course
in mathematics during their first year to introduce them to calculus and
statistics. The lecturer must be aware of conventions and terminology
in this area which differ from those elsewhere. A typical example is
the expression "marginal rate" which in mathematical terms is the first
derivative. Service courses to other areas of the social sciences tend
to be in the statistical and related areas.

Again a problem of motivation is likely to be encountered since the
majority of these students will only have an "O" Level mathematics entry
qualification.

SERVICE TEACHING TO THE ARTS

Service teaching of mathematics is not necessarily limited to areas
of study which depend upon numeracy. Students of Philosophy may require
a course in symbolic logic and students of History and Literature a
course on mathematics as a cultural influence. Such service courses are
not common but they do exist.

SERVICE TEACHING METHODS

Conventional lectures are the normal mode of teaching with students
being provided with additional written course material. The large
classes, which can exceed 200, are broken down into manageable tutorial
groups at which queries concerning examples are covered, together with
points raised by the lectures. For some courses more innovative methods
of teaching are employed, such as specially prepared television
programmes, self-paced learning, directed reading with discussions, and
some computer aided learning. This last depends upon good software, the
majority of which is still at the pilot study stage.

Assessment is normally by conventional examination although there
might be various pieces of work set throughout the year which are
collected in and marked as feedback to the student.

In general the service teaching of mathematics is centralised in
Departments of Mathematics although in some instances separate
Departments of Engineering Mathematics exist which service engineers.
It is in the interests of the Departments of Mathematics to provide a
good teaching service since approximately half of their staff and time
can be devoted to it. A poor response to a call for a service course
can result in the department requesting the service deciding to attempt
to teach it themselves. For this reason experienced lecturers are often
given important service courses to ensure that the customer is
satisfied. Help with tutorials can come from other mathematics staff,
research students, or staff from the department being serviced.

SUPPLY AND DEMAND

Some potentially serious problems are arising in Britain due to a
falling population at the age of 18 and an increasing industrial demand
for well qualified engineers, mathematicians and scientists. Various
estimates put this mismatch between supply and demand at about 7% by the
end of the decade and substantially worse by the end of the century.

Certain secondary problems are already apparent which will
exacerbate the situation. The shortage within industry of sufficiently
numerate and computer literate staff, combined with the relatively poor
pay of teachers in schools, has meant that teachers with the appropriate
skills can get better paid jobs in industry. The consequent shortage of
mathematics, computer science and science teachers within the secondary
school system is resulting in fewer students gaining sufficiently high
qualifications in these areas to obtain admission to tertiary education
in these subjects. Fewer still are likely to choose teaching as a
profession. An inevitable downward spiral appears to be developing.
The problem with respect to mathematics has been detailed by Howson (2).

This problem has been recognised at government level although there
is considerable discussion about appropriate action. Inevitably there
are financial implications. Two proposals which have been made are to
lower entrance standards and to retrain personnel under continuing
education schemes. In the former case students with lower standards in
mathematics would be accepted for engineering, mathematics and science
courses and the deficiencies corrected at the tertiary level. This
could lead to an extension of degree courses by one year. The
continuing education initiative involves admitting mature students with
appropriate aptitudes to retrain in shortage areas. A variation is to
encourage more women to consider this area of employment and to retrain
those so inclined, particularly any who may have left such an area to
raise a family.

One constant factor is the increasing requirement to teach
mathematics, not only to mathematicians, but to an expanding number of
other disciplines, as a service subject. This unglamourous and often
undervalued area of the tertiary educational system underpins the
maintenance of a strong scientific and industrial base. For this reason
alone sustained support for the service mathematician must continue to
be forthcoming.

REFERENCES

1. Statistical Bulletin 4/87, "International Statistical Comparisons in
    Higher Education", Department of Education and Science, London, ISSN
    0142-5013, March 1987.
2. Howson, A.G., "Challenges and Change", Inaugral Lecture, University
    of Southampton, January 1987.

MATHEMATICS AS A SERVICE SUBJECT IN ARGENTINA

N. Patetta
Universidad CAECE, Buenos Aires, Argentina

ABSTRACT

This report briefly describes the Argentinian educational system as a whole and, in more detail, the service teaching of mathematics in the country's Universities.

1. DESCRIPTION OF THE SCHOOLING SYSTEM

Education in Argentina is organized in three levels: primary, secondary and university. There are also some non-university post-secondary institutions, principally secondary teachers' training schools.

Primary education is a Provincial responsibility, but in all Provinces it is of seven years duration. There are some differences in overall syllabuses, but not in the mathematical topics which may be described as elementary arithmetic and geometry oriented to everyday life.

The secondary level is largely organized by the Provinces, but there are also National Secondary Schools depending directly on the Central Authorities. Both types of school cover five years of studies. With some regional differences, the mathematical syllabuses are about elementary (Euclidean) geometry, elementary algebra, trigonometry, and introductory topics of calculus. There are some schools specially oriented to different disciplines (accountancy, sciences, pedagogy, technical, etcetera) which introduce differences in the depth of their mathematical studies but not essentially in the topics. It should also be noted that this scheme exists in two subsystems: private and public. The private subsystem is supervised by the Central Authorities and in some cases a policy of subsidies is applied

2. THE UNIVERSITY LEVEL

There are twenty five National Universities, three Provincial Universities and twenty three private Universities in Argentina with a total of approximately 700,000 students (the population of Argentina is 30,000,000). The entrance system to the Universities is completely open, the only prerequisite condition being to have obtained a secondary certificate of studies from any secondary school. There is no special preparation for University entrance, though a few Universities offer non-compulsory introductory courses. Universities have autonomy so undergraduate programs vary substantially from University to University.

Practically all the Universities are organized in Faculties which involve similar courses of study. For instance Buenos Aires University (the most inportant one) has Faculties of Exact Science, Economic Science, Engineering, Agronomics, Medecine, etcetera. Each Faculty is organized in Departments which involve similar subjects. Each Faculty with enough mathematical courses has its own Mathematics Department in which mathematics is taught as a

service subject. The exception is, of course, the
Mathematics Department of the Faculty of Sciences, which
teaches service courses but also organizes the course of
studies for students of Mathematics. Obviously, with this
organisation, all of the teaching by the Mathematics
Departments is service teaching except in the Faculty of
Exact Sciences where service teaching is, generally, below
40% of the total. The situation outlined applies in most
Universities.

The kind of undergraduate mathematics service courses
offered in the different Faculties vary according to the
discipline, but generally they are courses of calculus,
elementary algebra, linear algebra, geometry (linear and
quadratic) and probability theory and statistics. Numerical
analysis is usually taught by the teachers of computer
science.

The Faculties with the most important Mathematics
Departments are the Engineering Faculties. In these, three
courses of calculus, one or two courses of algebra and
geometry and one or two courses of probability and
statistics are usually taught. The Faculties of Economic
Sciences have a similar structure but perhaps with only two
courses of calculus, these being at a slightly lower level.
It is interesting to notice the increase in the demand for
non-traditional courses by disciplines such as Biology, such
demand often being oriented to discrete mathematics and
modelling.

This organisation of service mathematics teaching,
allied with the very large numbers of students undertaking
University studies, has produced service Mathematics
Departments in which practically all the lecturers are part-
time instructors. In such Departments the only real
possibilities for research are "service research", ie
mathematical support for the technical research of the
specific Departments. Usually such research is in the areas
of numerical analysis, ordinary differential equations,
partial differential equations and similar fields.

The fact that mathematics courses are usually the first
courses of any course of studies, allied with the
unrestricted nature of admissions to Universities, naturally
makes mathematics act as a "filter" for the user
Departments. This causes a serious problem for mathematics
teachers because the students often have the feeling that
they are having artificial difficulties produced by subjects
which are unimportant for the discipline which they really
want to study.

Each University faces its own problems with different
ideas and in a different regional context. The brief
description just given is a very general one and, obviously,

it is not the same for Buenos Aires University (200,000
students) as for the smaller Universities in the Provinces
(10,000 students or less). The smaller number of students
simplifies the problem, and it is possible to have a more
informal organization. In such Universities the different
Faculties share a single Mathematics Department which can
have a more professional teaching body. Teachers in these
Departments can alternate service teaching with research and
it is possible to have a staff of full-time lecturers within
a restricted budget which would be impossible in the big
Universities.

SERVICE MATHEMATICS IN AUSTRALIAN HIGHER EDUCATION

A.G. Shannon, J.G. Sekhon
New South Wales Institute of Technology, Broadway, Australia

ABSTRACT

This paper briefly describes the Australian education system particularly the ways in which it affects recruitment to tertiary education institutions. The position of service mathematics courses in tertiary education is then described in more depth.

1. INTRODUCTION

1.1 Motivation for this study

The question of who should teach mathematics to non-mathematics majors has taken on a new urgency in a time of funding cutbacks. In a steady-state staffing situation, it is not unreasonable that serviced departments should take a second look at retention rates and applicability of content in service mathematics courses. Co-operation is essential and it can only be fruitful if peripheral issues are removed from the demain of discussion. The motivation for some of this study also came from a concern with an increasing tendency for the proliferation of specialist first-year mathematics courses at the expense of core material. This splintering of first-year core mathematics has frequently arisen because of high failure rates in mathematics by students who will need some mathematical tools and a mathematical conceptual framework in other disciplines. This is not totally unexpected, given the growth of quantitative methodology in a variety of areas, together with an increasing percentage of the population in higher education.

1.2 Education practices in Australia

Educational practices are the responsibilities of the six States of Australia , though there are Commonwealth (Federal) committees which disburse money to them and to the two Territories. The Commonwealth still provides approximately one-sixth of funding directly for schools, over a quarter of funds for Technical and Further Education (TAFE), and all the funds for universities and colleges [1].

There have always been tensions and duplications of bureaucracies between the States and the Commonwealth governments [2]. The late eighties are no exception with changing demographic patterns and increasing strains on the Federal government's social services safety net. Educational structures vary from State to State. For example in New South Wales primary education runs to year 6 whereas in Queensland it runs to year 8. Hence, it would unnecessarily complicate issues to give a detailed analysis. In general, students matriculate to university at the end of Year 12 at age 18.

There have been many reports of commissions of enquiry into various facets of Australian education in recent years. In so far as one can generalize, the proposals are moving towards general education up to Year 10, the period of compulsory education, followed by specialization of sorts for Years 11 and 12. By this is meant that alongside the traditional academic curriculum at this level would be a variety of alternatives, including TAFE courses to give more

subject choices [3]. Enrolments in TAFE are expected to
increase by about 50% between 1984 and 1990, partly due to
these increasing cross-sectoral developments between schools
and other sectors of tertiary education [4].

More Australian school students are completing
secondary school than ever before. The statistics indicate
that Australia-wide in 1986, 48.7% of students went on to
Year 12 compared with 46.6% in 1985 and 36.3% in 1982 [5].
The trend has been welcomed by government, community and
business organisations and trade unions. Historically,
Australia has recorded low secondary education participation
rates compared with OECD countries which constitute our
major trading partners. Likewise participation in higher
education in Australia has been comparably low, and surveys
by the Australian Vice-Chancellors' Committee in 1985 and
1986 have shown that many thousands of qualified students
are unsuccessful in gaining a place in higher education. At
the same time, the Commonwealth Government has a stated aim
of achieving an overall Year 12 participation rate of 65% by
1990.

1.3 Australian higher education
Not surprisingly, the extent and depth of school
mathematics varies considerably from State to State and even
within some States. In practice, there is not much movement
by students across State boundaries, and universities tend
not to set prerequisite content material in mathematics.

Selection to entry into Australian universities and
colleges of advanced education (CAEs) varies among the six
States and two Territories. All systems provide for the
admission of mature age students, but the greatest variation
occurs in the selection procedures for those who are
completing their secondary schooling (usually at the end of
Year 12). About 10% of the 17-24 year old population cohort
participates in higher education [6]. A further
complicating factor is that most institutions of higher
education operate quota systems for admission to the
different discplines ranging from medicine and veterinary
science near the top to arts and science generally near the
bottom.

"Systems differ in the proportion of a student's work
in the final years of secondary education which is
considered in determining admission to higher education.
Some systems use assessments in all subjects taken and thus
oblige students seeking admission to choose all their
subjects from the category that may count. Other systems
use only a proportion of the subjects studied to obtain
assessment of performance for admissions purposes and thus
leave students some freedom of choice in their subject
enrolments." [7]

All universities and colleges admit some students who
do not meet normal entry requirements and have been
disadvantaged but only if there is a good chance that they
will be able to cope with their studies [8]. Generally
though they are "closed" systems of entry and overall
graduation rates are quite high. Success rates (ratio of
subjects passed in a given year to the total number of
subjects enrolled in that year) vary from a high of 94% in
Medicine and Dentistry through 85% in Engineering to 78% in
Arts, Commerce and Economics [9].

Failure rates can be quite high in first year service
mathematics courses which sometimes serve as filters for the
user departments. Lack of relevance and inadequate
secondary school preparation are common cries from different
sides. One suspects too that some of the quantitative work
may appear too early in the degree programs simply because
someone is going to refer, however obliquely, to some
technique.

There is also pressure from the accrediting authorities
which are of two kinds: government and professional. The
former give an institution the right to award degrees at
different levels, but the latter exercise the control over
the graduate to practise a profession. We cannot but agree
with James [10] of the need for mathematics educators to
acquaint themselves with those areas of mathematics used in
industry and business and of the employers' expectations of
new graduates' skills. Then one is in a better position to
negotiate with the professional society about what
constitutes a core mathematics curriculum to develop the
conceptual framework for the profession and the role of the
cultural heritage in the education of the professional.

2.   SURVEYS OF USERS

2.1 Staff and student expectations

In recent years little attention seems to have been
paid to the expectations of students enrolled in tertiary-
level mathematics courses. Shannon and Sleet [11] report a
study, carried out at The New South Wales Institute of
Technology, which investigated first-year students' opinions
of the aims of mathematical courses and compares their views
with those of lecturers. We believe a knowledge of these
feelings to be an important consideration in designing
mathematics courses.

A list of 33 possible aims for any undergraduate
mathematics course was drawn up by the authors after
consulting other members of the mathematics staff, and after
perusing other reference sources. These aims related to the

needs of students who were studying mathematical courses.
They were not necessarily mathematics aims exclusively.
        The aims were listed randomly and first given to the
relevant staff in a questionnaire. The purposes of the
staff survey were to give content validity to the subsequent
student survey, and to obtain staff views of the listed
aims. Staff were asked to comment on any aims which they
would like to see added. The students' survey, which was
administered in the third week of their first year, looked
at the extent to which students' perceptions of aims and
expectations depend on the discipline they are studying.
Both surveys were anonymous. Whereas the staff were asked
to rate the aims according to whether they considered them
to be either important for their discipline, of general
importance only, or of no importance at all, the students
were asked to rate the same aims merely as important,
unimportant or incidental.
        Students and staff in the disciplines of science,
engineering, business studies, computing and mathematics
were surveyed. Out of 200 possible staff, 68 replied, as
did 959 out of 1050 students. Some of the students who
completed the questionnaire were subsequently interviewed
individually to check on their understanding of the aims.
        The aims which all staff consider important are
concerned not only with relating mathematics to, or showing
the applications of mathematics in, industry, research
investigations, probability, statistical inference and other
subjects, but also with students' acquiring an ability to
construct a mathematical model and to deduce relationships
from experimental data. Staff in all disciplines have the
view that the applications of mathematics are important.
The study also shows that students from all discplines also
see these aims as important. In addition, they are
concerned about learning mathematical principles, and
developing an ability not only to solve mathematical
problems but also to think logically through a study of
mathematics. The responses to some aims, however, implied
that students in science, engineering, business studies and
computing are studying the subject merely to obtain a degree
in their chosen discipline. It appears, therefore, that
they have indicated a desire to learn mathematical
principles and problem solving methods not because they are
interested in the subject, but because they consider it
necessary to achieve such aims in order to pass examinations
and obtain a degree.
        Details can be found in the published paper. Suffice
it to say that a study such as this can contribute to a
climate of cooperation between mathematics and other staff

and help the mathematics staff to be more aware of what
motivates staff and students in the user departments.

2.2   Who teaches service mathematics?

      To ascertain trends and views on this topic, a
questionnaire was sent to mathematics departments in 66
Australian universities and colleges in October 1982, and 40
were returned in the specified time [12].

      Of the 33 mathematics departments responsible for
teaching mathematics to students enrolled in other courses,
some 25 (75%) devote more than 40% of their total teaching
commitment to the teaching of service subjects; and 13
(39%) of these departments give the responsibility of the
service teaching to full-time staff members who have special
expertise and/or interest in teaching service mathematics.
Of the 40 responses, 34 (85%) considered that mathematics
subjects for non-mathematicians should be the
responsibility, totally or mainly, of the mathematics
department of the institution concerned.

      In 21 institutions (53%) there have been moves by
"serviced departments" to have the mathematics subjects of
their courses taught by their own staff members. In 15
(71%) of these serviced departments, if the moves had been
successful, the staff members undertaking the teaching would
not be mathematicians specially recruited by that
department.

      A clearly defined institutional policy on service
teaching (not only for service mathematics) exists in 13
(33%) of the tertiary institutions whose Head of Mathematics
Department responded to the questionnaire. Comments on
moves by serviced departments to teach their own mathematics
subjects include discussions on serviced departments'
wishing to integrate the mathematics of their course
structures into those subjects of the course where the
mathematics is "used", and mathematics departments offering
the first year subjects but not those required in later
years. General comments related to concern about a
developing trend for some social science departments to
drop, or make voluntary rather than essential, mathematics
and/or statistics subjects so that courses become more
popular with marginally competent students. This is not in
the best interests of the more able students in those
courses.

      A need for greater liaison between serviced and
servicing departments with more discussion on joint teaching
projects to complement the efforts being put in by some
mathematics departments in this area was also mentioned, as
were the dangers associated with wrong conclusions which can
arise when non-mathematics staff lead students through
incorrect mathematical analysis of a problem. Many

mathematics departments already have the mathematicians on staff who have expertise in areas of engineering, the physical sciences and management science. In general, mathematics departments would seem to be best equipped to help students see mathematics as a whole and to illustrate a wide range of applications of particular concepts.

There has been a steady decline in mathematics service teaching during the last decade, a trend which has been increased by the funding arrangements of the last few years, despite the growth in the numbers of new areas which use mathematics. A second survey [13] of user departments tended to confirm that there is, as there has been for some time, a trend in many Australian tertiary institutions for the service subjects traditionally offered by mathematics departments to be critically re-appraised by the serviced departments. This appraisal includes moves to remove the responsibility for these subjects and their mathematical content from the mathematics department to the "user" department. Such a trend if continued, poses a serious educational problem for the students concerned. There is reason to believe that the mathematical education of these students could be lacking in that they would view mathematics (and statistics) in the narrow "technique oriented" framework in which it is so often presented.

2.3 What mathematics to teach

Results discussed here are based on a number of surveys undertaken during the period 1977-84. These surveys have looked at the syllabuses taught and the opinions of employers. The description here of the present position is in broad terms only, as the differences that exist between one institution and another and between one branch of engineering and another are not inconsiderable.

Table 1 gives the main colleges of advanced education in the various Australian States that provide engineering courses and the minimum length of their full-time study in years. The column giving the total number of hours of mathematics, as a percentage of the total course time, is broadly indicative of the hours (averaged for the three disciplines: civil, mechanical and electrical) time-tabled in the main colleges of advanced education. Two universities have also been included for comparison [14].

Eng. Course at	Min. Length of Course (years)	% Mathematics (Avg. of Courses)
University of Sydney	4 FT	modular degree
University of New South Wales	4 FT	11.1
NSW Institute of Technology	6 academic years (PT or sandwich)	12.5
Royal Melb. Inst. of Tech.	4 FT	12.7
Chisholm Inst. of Tech.	4 FT	12.2
Swinburne Inst. of Tech.	4 FT	14.4
Qld Inst. of Tech.	4 FT	9.2
S.A. Inst. of Tech.	4 FT	12.8
Curtin Univ. of Tech. (WAIT)	4 FT	12.3

Table 1, Duration of courses and allocations of hours

Electrical engineering courses, both at universities and the CAEs, tend to cover more mathematics than others; civil engineering courses have generally the least time devoted to mathematics courses.   The content of the courses is traditional and conservative. Bajpai and Singh [15] list typical mathematical topics. The major CAEs offer them in a co-operative (sandwich) mode as well as part-time.

Most institutions which prepare students for first degree qualifications in engineering aim to give a deeper and broader understanding of fundamentals, a solid base for the technical science of their speciality. There is a conformity of pattern between courses in various institutions in the earlier undergraduate years. This is not necessarily to say that the courses, which are similar in content, reach the same standard.

There have been many changes in the patterns of higher education such as the development of various forms of credit-systems or modular schemes. Although considerable differences with respect to required versus elective courses exist, most engineering programmes allow students some freedom in mapping their own programmes of studies from a wide range of possible electives. At the upper undergraduate and graduate level, it is not uncommon for institutions to provide flexible schemes of study, whereby students may match their education to their abilities, aptitudes and career aspirations by selecting courses from a variety of mathematical and technical electives. The more advanced topics in mathematics are usually presented as optional subjects in the later stages of the degree course.

A questionnaire was also sent to PhD graduate mathematicians, engineers and scientists working in Australian industry to seek their comments on specific areas of mathematics of particular relevance to industry. To this

end, respondents were asked to consider each of 26 subject
areas from the standpoint of frequency of usage and degree
of importance. Frequency and importance were used as scales
to measure the graduates' perceptions of essentiality of
mathematical methods [16].

According to respondents, mathematics has its part to
play in industry, but other elements enter, and cases abound
in which these other elements are overwhelmingly more
important than the precision of mathematics per se. As one
respondent informed us: "I view with some reserve
suggestions which emanate from industry and academia as to
just what sections of mathematics are important and,
therefore, must be included in university courses.
Industry's use of mathematics is not limited by non-
availability of knowledge of particular techniques as by the
difficulty of bringing problems and techniques together.
Therefore, whatever sections of mathematics are taught, it
is very important for university teachers of mathematics,
and indeed for all teachers, to have in mind that, in the
world at large, the identification of key problems is often
more difficult and just as important as their solution.
Ability to establish the right questions is more valued than
ability to solve them, and ability to extract the problem
for which equations should be written may be valued more
highly still".

There was a heavy emphasis on discrete mathematics and
statistical computing. Although certain areas of
mathematics scored lowly on frequency and importance scales
one should not automatically use relevance as the only, or
even principal, criterion for curriculum development in
service mathematics.

3. CONCLUDING COMMENTS

3.1 Service teaching
Service teaching is not always highly regarded in
Australian higher education. Lublin [17], in deploring the
low status given to teaching in general by comparison with
research in Australian universities, comments on the
traditional separateness which continues to be maintained
between disciplines and the on-going debates over service
subjects. "Is maths to be taught to all comers as the prime
discipline it is, or should, for example, architecture
students get a specially designed course in architectural
mathematics? ... How can you maintain motivation in first
year engineering students when 75% of their course work is
taken outside the Engineering Faculty?" She goes on to
mention how students compartmentalise their learning and
points out how part of the solution is in our own hands: to

integrate learning in a real life problem solving way. One
of the best examples of this is the medical degree at the
University of Newcastle, NSW, in which the course is built
around solving the problems associated with particular
diseases, rather than the traditional subjects of anatomy,
physiology, and so on [18].

3.2   Continuing education
     Sekhon [19] argues the inevitability of increased
graduate education to reduce educational and human
obsolescence and discusses inter-disciplinarity, synthesis
and mathematical modelling. In particular, he addresses the
ends of continuing mathematical education. This is being
done in Australia by mathematicians working in partnership
with companies and professional societies by putting on
tailor-made short courses which run for a few days on a
specific topic. Many universities and colleges in Australia
now have their own research and development companies which
can provide academic mathematicians with opportunities to do
consulting work in industry and commerce. NSWIT's R and D
company is called Insearch and also provides continuing
education for industrial companies. These are ways for
mathematicians to become more aware of some of the
environments in which the students they teach in service
courses will eventually work. Another way to do this is to
spend an occasional sabbatical leave in industry. Not
everyone needs to be involved in this way, of course, but
without some involvement by some staff the employers'
perceptions of undergraduate education can be very negative.
For instance, recent surveys for the Business Council of
Australia and of the 500 most profitable businesses in
Australia have indicated that more than 40% of respondents
regard tertiary educational institutions as lagging behind
current industrial/business practices [20].
     Mathematicians also contribute to continuing education
through such adult education organisations as the Workers
Education Association. These range from general interest
courses to bridging courses which provide a second-chance
for many adults to re-enter the formal educational world.
Increasingly, universities and colleges are becoming more
aware of their responsibilities, and of the opportunities,
in recurrent education [21].
     This paper attempts to do no more than set out some of
the questions (and findings) arising from our research. The
very fact of their being asked, however, must surely be a
positive step forward in the field of curriculum development
in the area of service mathematics - particularly at a time
when higher education is repeatedly urged to respond to the
manpower needs of industry and the professions.

REFERENCES

1. K.G.Mortensen, <u>Politics and Sociology of Funding</u>
    <u>Australian Schools:1962-84</u>, Gerald Griffin Press,
    Parkville, (Vic), 1985, Ch 10.
2. P.D.Tannock, <u>The Government of Education in Australia</u>,
    University of Western Australia Press, Nedlands, 1975,
    Ch 2.
3. G.W.Basssett, The future of education in Queensland,
    <u>Newsletter of The Australian College of Education</u>, 5(4),
    1986, p 3.
4. R.Broadbent, Crucial role for TAFE, <u>Newsletter of the</u>
    <u>Australian College of Education,</u> 5(3), 1986, p 5.
5. A.Thomas (ed.), Big rise in Year 12 retention, <u>Journal</u>
    <u>of Advanced Education</u>, 10(1), 1987, p 5.
6. Commonwealth Tertiary Education Commission, <u>Review of</u>
    <u>Efficiency and Effectiveness in Higher Education</u>,
    Australian Government Publishing Service, Canberra,
    1986, p 86.
7. B.McGaw, Selection of students for higher education,
    <u>Unicorn:Journal of the Australian College of Education</u>,
    13(1), 1987, pp 4-9.
8. Committee of Enquiry into Education and Training,
    <u>Education, Training and Employment</u>, Vol.1, Australian
    Government Publishing Service, Canberra, 1979, p 167.
9. Australian Vice-Chancellors' Committee, <u>Survey of</u>
    <u>Student Progress 1982</u>, April, 1986.
10. D.J.G.James, The usage of mathematics in the IT
    industry, <u>Bulletin of the Institute of Mathematics and</u>
    <u>Its Applications</u>, 22(9/10), 1986, p 130.
11. A.G.Shannon and R.J.Sleet, Staff and student
    expectations of some undergraduate mathematics courses,
    <u>International Journal of Mathematical Education in</u>
    <u>Science and Technology</u>, 9(2), 1978, pp 239-247.
12. M.L.Fuller and A.G.Shannon, Service mathematics: a
    sevice to whom? <u>Australian Mathematical Society</u>
    <u>Gazette</u>, 10(4), 1983, pp 77-81.
13. M.L.Fuller and A.G.Shannon, The teaching of service
    mathematics in undergraduate courses - a cause for
    concern?, Fifth International Congress on Mathematical
    Education, <u>Abstracts of Short Communications</u>, 1. Oral
    Presentations, 1984, p 122.
14. J.G.Sekhon and A.G.Shannon, Mathematical education of
    engineers: an Australian perspective, <u>European Journal</u>
    <u>of Engineering Education</u>, 10(3/4), 1985, pp 295-303.
15. A.C.Bajpai and J.G.Singh, A survey of mathematics
    courses for engineers in Australia, <u>Tertiary Education</u>
    <u>in Applied Mathematics Project</u>, Report No. 15, NSWIT,
    Sydney.

16. J.G.Sekhon, <u>The PhD Education of Industrial Mathematicians in Australia</u>, Unpublished PhD Thesis, University of New England, 1985.
17. J.Lublin, Teaching and learning in higher education - promise and performance, in R. Philps and A.G. Shannon (eds), <u>What Do We Expect of Education</u>? The Australian College of Education, Carlton (Vic), 1983, pp 107-116.
18. D.Maddison, Educational and assessment strategies, <u>The University of Newcastle Faculty of Medicine</u>, Working Paper No.1, 1975.
19. J.G.Sekhon, Technological change and the mathematical education of engineers: challenge and response, <u>International Journal of Mathematical Education in Science and Technology</u>, 15(2), 1984, pp 153-160.
20. R.Millikan, Education fails the test of business, <u>Newsletter of the Australian College of Education</u>, 5(3), 1986, p 1.
21. J G.Sekhon and A.G.Shannon, Recurrent and Second Chance: Two Roles for Colleges of Advanced Education in Continuing Education, <u>Australian Journal of Adult Education</u>, 21(2), 1981, pp 3-9.

MATHEMATIQUES ET DISCIPLINE DE SERVICE
LE POINT DE VUE D'UN INGENIEUR

G. Aillaud
I.B.M., Paris, France

Votre texte de consultation sur l'enseignement des mathématiques m'a vivement intéressé et c'est au titre de mon expérience professionnelle que je me permets de participer à ce débat.

Je ne suis ni mathématicien ni enseignant. Par contre 24 années de travail comme ingénieur dans une entreprise d'informatique m'ont toujours mis à cheval entre les deux :
- 2 ans de modélisation mathématique dans un service de prévisions et de planification,
- 10 ans de recherche appliquée en combinatoire, théorie des graphes, recherche opérationnelle, avec deux activités principales :

1. résolution de problèmes concrets, recherche d'algorithmes,
2. éducation pour ingénieurs confirmés (théorie des graphes),

- puis 12 ans de marketing, dont 2 dans un domaine industriel particulier (la chimie) et 10 recouvrant un ensemble étendu d'applications scientifiques -dont l'intelligence artificielle et les systèmes experts depuis 2 ans et demi- toujours orienté vers le conseil en résolution de problèmes, l'éducation et la formation d'ingénieurs.

Je voudrais citer ici le nom de M. Jean-Claude HERZ qui sut m'initier aux rapports entre la théorie et la pratique lors de mes 10 ans de recherche appliquée et à qui je dois beaucoup d'idées exprimées ci-après , même si je les présente à travers une expérience personnalisée.

Quelques considérations générales.

Je pense qu'il existe de "grandes familles" d'outils et de disciplines mathématiques relevant de formes de raisonnement différentes.
Je crois que les démarches
- analytique en analyse numérique,
- probabiliste en statistiques,
- souvent heuristique en combinatoire,
- intuitive en géométrie,
ne relèvent pas des mêmes processus de pensée. Les "habitudes" à acquérir, l'expérience nécessaire n'y sont pas identiques.

J'ai par ailleurs remarqué -et les problèmes actuels de "mobilité"
accentuent l'importance de cette remarque- qu'un statisticien, par
exemple, apprendra *facilement* les techniques de simulation discrète
(processus markoviens, modélisation d'évènements aléatoires, interprétation de résultats sur des files d'attente, etc...), tandis qu'un spécialiste de géométrie (et de calculs 3D dans un bureau d'études) aura plus
de difficulté à apprendre cette même simulation discrète à 40 ans.

De la même façon, un ingénieur français de 50 ans (n'ayant donc pas
appris l'algèbre linéaire lors de ses études) a beaucoup de mal à
pénétrer le domaine de l'analyse des données multidimensionnelles
(calcul matriciel, vecteurs propres et valeurs prores...).

Cela implique deux conséquences.

1ère conséquence
Il faut que les entreprises l'admettent et que la "carrière" puisse
se dérouler autour d'une *famille de disciplines* formant *la base d'une
spécialité.*

La *"mobilité"* ne peut s'exercer n'importe comment, sous peine de
graves gâchis.

Il est souhaitable d'être mobile en *approfondissant* les connaissances de base de *sa* famille de disciplines (le métier) tout en
élargissant celles des *domaines d'application.*

Exemple : un spécialiste de combinatoire peut avantageusement
passer de problèmes d'emplois du temps à des problèmes de découpe,
d'ordonnancement, de tournées, etc... pour devenir spécialiste de
problèmes d'affectation, empiétant progressivement sur toute la recherche opérationnelle, la programmation mathématique, les systèmes experts,
etc... A l'inverse, passer de la combinatoire à l'analyse numérique à
45 ans aura 9 chances sur 10 d'être un gâchis.

2ème conséquence
L'enseignement supérieur doit tenir compte de ces différentes
formes de pensées, sans pour autant "tout" enseigner à chacun : l'acquisition de connaissances ne s'arrête pas à la porte de l'université.

Il convient donc d'affiner les *critères de choix* de ce que l'on va
décider d'enseigner (et de ne pas enseigner) dans un cursus
d'ingénieur.

Il convient ensuite d'appliquer les critères choisis aux différentes disciplines.

Je n'ai pas l'ambition de résoudre ce problème, mais simplement de développer trois conditions que je pense être nécessaires (et suffisantes???) pour qu'un ingénieur puisse utiliser, dans sa vie professionnelle, les outils mathématiques dont il peut avoir besoin.

1ère condition : qu'il connaisse l'existence des outils disponibles.

Elémentaire, mon cher Watson ?

Pas si élémentaire que cela : il lui faudra beaucoup lire ; mais l'ingénieur n'est pas "habitué" à lire des textes mathématiques nouveaux et le temps lui est compté.

D'où des conditions nécessaires pour pouvoir réaliser ce voeu.

1ère nécessité : que la "culture" mathématique acquise lors des études permettent de *lire aisément* des articles ou livres de *bon niveau*. Cela nous ramène à l'introduction : il lui aura fallu, durant son cursus universitaire, "balayer" différentes formes de raisonnement et de pensée, au risque, plus tard, de se voir confiné aux seules qu'il aurait apprises.
Le rôle de l'enseignement est ici de dispenser *ce qu'il serait trop difficile d'apprendre seul* (concepts et modes de raisonnements plutôt que des sommes de théorèmes).

2ème nécessité : que les écrits s'adressant aux ingénieurs soient à leur portée. Ceci me paraît capital.
Des livres à l'usage d'étudiants ou d'universitaires peuvent, en effet, être denses, voire ésotériques. A l'inverse, l'ingénieur qui passe 90 % de son temps à d'autres activités ne peut se permettre des lectures au rythme de deux pages à l'heure ! il ne lira pas Bourbaki. Il aura beaucoup de mal à lire certains livres actuels.
Il est alors indispensable que les "savants", les universitaires, écrivent à l'usage des ingénieurs. Ce n'est pas de la vulgarisation grand public. On s'adresse ici à un public ayant un bagage mathématique. Appelons cela de la vulgarisation au 2ème degré.
A titre d'exemple, Matridakis en prévisions, Berthier et Bouroche en analyse des données multidimensionnelles... les pionniers : Kaufmann en combinatoire ou Motles en statistiques sont des réussites.
Il faut une panoplie de livres *riche* et fréquemment *mise à jour*. Donc également des crédits. Cela pourrait relever d'une coopération entre Education Nationale et Industrie.

2ème condition : qu'il comprenne la signification des outils disponibles.

Elémentaire, mon cher Watson ?

Oh que non ! Combien de fois ai-je vu des problèmes d'optimisation -du diamètre d'un tuyau dans un processus continu par exemple- que l'on pouvait calculer par une approche

- en simulation continue (processus continu = équations différentielles...)
- ou en simulation discrète, car on pouvait modéliser en discré-
tisant sans risque d'erreur,
- ou en programmation linéaire, car on pouvait linéariser par
morceaux aussi bien les contraintes que la fonction économique,
- ou enfin par des heuristiques, la modélisation mathématique pouvant être lourde et l'approximation "à la louche" largement suffisante,

et personne ne savait choisir une approche plutôt qu'une autre, dans
l'ignorance de la signification de chaque famille d'outils.

Autre exemple, plus fréquent, l'analyse des données multidimensionnelles. Quel ingénieur sait-il choisir entre une analyse factorielle
en composantes principales plutôt qu'en correspondances ? Entre une
analyse discriminante, canonique, etc... ?
Et même : pourquoi une approche géométrique de recherche de
propriétés, comme ci-dessus, plutôt qu'une approche de classification ?
Quel ingénieur connait-il la signification des métriques utilisées ?

Or le choix des méthodes aura un impact sur la *signification* des
résultats.

Pour répondre, il faut avoir intégré la *signification* des méthodes
utilisées. Il faut avoir lu et nous ne répéterons pas ce qui a été dit
plus haut.

Nous ajouterons cependant que les publications de vulgarisation au
2ème degré doivent être *bourrées d'exemples vécus*. En effet, le passage
de la formulation d'un problème à sa formalisation -ou, si l'on
préfère, du cahier des charges à la modélisation- ne s'enseigne guère
actuellement ; la seule réponse réside dans l'expérience, l'habitude,
l'accumulation d'exemples.

3ème condition : qu'il sache interpréter ses résultats.

C'est la condition principale et, de loin, la plus difficile à
réaliser. Condition principale car "l'utilisateur" d'un outil est le
SEUL RESPONSABLE de l'interprétation de SES résultats.

Un ingénieur d'aciérie pourra toujours se faire AIDER par un
statisticien -de l'entreprise ou extérieur à l'entreprise- pour une
modélisation un peu "pointue", mais il reste le seul à connaître son
domaine d'application et le statisticien ne l'aidera guère dans une
interprétation *concrète*.

Et c'est pourquoi notre cher Watson ne peut se permettre de considérer comme élémentaire la 2ème condition. Et cela donne encore plus
d'importance aux livres de vulgarisation au 2ème degré agrémentés d'une
multitude *d'exemples interprétés*.

Les exemples doivent illustrer, bien sûr, les cas généraux, mais
également des cas inhabituels, "limites". (Notons ici que l'ingénieur
est bien préparé, par sa pratique, à l'ambiance de l'inhabituel.

L'automatisation l'habitue à travailler "par exception", c'est-à-dire lorsque le processus automatisé est en disfonctionnement, les cas habituels, banals étant traités par le robot, le système expert, etc...).

En d'autres termes, la "couverture" du domaine d'application d'un outil doit être assez complète dans une vulgarisation au 2ème degré, d'où ma proposition de coopération Université-Industrie (auxquels il serait bon en France, d'ajouter les Sociétés de Services qui sont souvent d'ores et déjà un bon exemple d'une telle coopération).

Permettez-moi d'ajouter deux remarques à cette -déjà longue- lettre.

1. Je regrette de n'avoir rien lu, dans votre article, sur la *formation continue* post-scolaire.

Il serait intéressant de se demander quelles sont les conditions pour qu'elle soit *possible* et *efficace*.

Rejoignant ce qui précède : quelles sont les acquisitions initiales nécessaires (concepts, "culture", réflexes...) et quelles sont ensuite les lectures et les enseignements à conseiller : contenu, charge de travail ou durée, fréquence, modalités concrètes.

2. Vous mentionnez, dans votre texte, que d'aucuns préconisent l'intervention d'ingénieurs dans certains enseignements supérieurs. Je pense que ce serait bon, à la condition de ne pas être brutalement massif et de rester sous le contrôle exclusif de l'Education Nationale, des Universités et des Ecoles.

Je pense qu'il serait *très* intéressant de se poser la même question pour l'enseignement secondaire et, dans un premier temps, pour les classes de 3ème, 2nde et 1ère du cursus français. Cela nécessite bien entendu une formation pédagogique complémentaire pour les ingénieurs volontaires (et exclusivement volontaires).

On peut envisager des détachements à temps plein pour une période brève (3 ans) ou des détachements à temps partiel pour des durées plus longues (tiers de temps ou mi-temps sur 5 ou 6 ans ?).

Nombreux seraient les ingénieurs expérimentés volontaires. Il s'agit de gens capables, parfaitement susceptibles de "faire aimer les maths" à une population d'adolescents. N'est-ce pas un des problèmes clefs de l'enseignement secondaire ? Les ingénieurs sont une mise pédagogique de premier ordre.

Bien entendu, le processus inverse, plus connu, est indispensable : que les professeurs du secondaire fassent des stages prolongés dans l'industrie, sur des problèmes réels, cadences et modalités à définir très souplement.

Je terminerai par une boutade.

L'arrivée des calculettes dans le secondaire il y a 10 ans

nécessite des modifications pédagogiques qui ne sont pas encore entièrement maîtrisées. Imaginons-nous ce qui va se passer, dans un proche futur, dans le secondaire et le supérieur, lorsque le *calcul formel* se traitera sur des outils grand public du même type ?

... Ce sera une autre paire de manches !

INTUITION GEOMETRIQUE — INTUITION PHYSIQUE

G. Châtelet
Université de Paris VIII, Paris, France

ABSTRACT

Geometric intuition, when compared to calculus, is often considered as a mere "illustration". We grant geometry for being a source of visual evidence, for global understanding but suspect it to be less rigorous.

The situation is worse as far as Mathematics and Physics are concerned : poor epistemology generally considers Physics as "dirty mathematics" and mathematical objects as "tools for physicists" committed to produce "models".

Through a brief analysis of some prominent scientific events of the early XIXth century : invention of the Argand plane, Hamilton's theory of quaternions, Galois theory of equations, Ampère-Faraday-Maxwell electromagnetic theory, we sketch another point of view ; we try to understand what we suggest to call modern physico-mathematics (<u>very different from the so-called Mathematical Physics</u> !).

Classical geometry was devoted to the study of figures and distance ratios of real objets.

"Abstract" algebraic operations and geometric intuition remained separated. With the Argand-Wessel plane and quaternionic geometry, the very constitution of time and space became an object of mathematical investigations. Following Hamilton we have to "regard Algebra as being no mere Art, nor language, nor primarily a Science of Quantity but rather as <u>Science of order in progression</u>". J. C. Maxwell perfectly understood this revolution and "this mode of contemplating geometrical and physical quantities more primitive and more natural than the other."

Combined with Faraday's "lines of force", Maxwell succeeded in his attempt to create a true <u>electrogeometric space</u>.

In an apparently very different domain, Galois theory of equations
does not urge to find formulaes <u>giving</u> the roots (x = ?) but describes
the very process of discerning the roots. Geometry, Physics, Algebra do
not want only to <u>identify individuals</u> any more, but try to <u>produce</u> them
and a systematic use of <u>diagramms</u> is the specific language of this triple
alliance. Physico-mathematical diagramms possess the great creative power
of sketching the very acts of getting acquainted with some continuum.

Le lecteur moderne, toujours un peu pressé, serait probablement
perplexe et agacé par les "Lectures on Quaternions" de Hamilton. Il
pesterait certainement en tous cas contre les cinquante pages consacrées
à la définition des vecteurs de $\mathbb{R}^3$. Un de nos excellents ouvrages
d'algèbre n'écrit-il pas que : "L'espace-temps n'est que l'espace $\mathbb{R}^4$
muni de la métrique (-+++). Pourquoi faire tant d'histoires ?"

"Notre" espace ne serait donc qu'un "misérable cas particulier" de
$\mathbb{R}^n$. Comprendre l'espace serait simplement "abstraire", "généraliser" et
bien sûr "formaliser"! L'opinion courante brode souvent sur le thème :
"la géométrie illustre les calculs algébriques - on voit mieux mais sa
rigueur laisse à désirer" ou "la physique n'est qu'une cuisine sordide,
une mathématique mal dégrossie"[1]. La mathématique est toujours la reine
des sciences (l'arithmétique en étant le plus précieux joyau), mais on
exige paradoxalement qu'elle se rende "utile", qu'elle troque à volonté
son sceptre contre un tablier de bonne à tout faire.

Nous espérons ici, à l'aide de quelques exemples contribuer à la
réflexion sur des questions comme "l'application de la mathématique à la
physique est-elle providentielle?" Cette application est-elle
nécessairement un rapport de subordination ? Quels sont les rapports
entre intuition géométrique, calcul algébrique et sens "physique" ?

Mentionnons tout d'abord un fait historique surprenant :
l'identification des nombres imaginaires au plan n'apparaît dans la
littérature mathématique qu'au début du XIXe siècle[2]. La synthèse des
gestes géométrique effectués dans le continuum à deux dimensions et des
calculs impliquant les "quantités impossibles" a donc demandé presque
trois siècles !

Les mathématiques du XVIIIe siècle avaient pourtant déjà abordé des
problèmes beaucoup plus compliqués ! La frontière qui séparait alors
intuition spatiale et entendement déductif délimitait strictement les

domaines de l'algèbre (la science des magnitudes) et de la géométrie (la science des figures).

Leibniz n'appréciait guère ce divorce et avait pressenti la possibilité de calculer sur les "rapports de situations" (Analysis Situs) comme en témoigne sa lettre à C. Huyghens : "J'ai trouvé quelques éléments d'une nouvelle caractéristique tout à fait différente de l'algèbre, et qui aura de grands avantages pour représenter à l'esprit exactement et au naturel, quoique sans figures, tout ce qui dépend de l'imagination. L'algèbre n'est autre chose que la caractéristique des nombres indéterminés et des grandeurs. Mais elle n'exprime pas directement la situation, les angles et le mouvement, d'où vient qu'il est souvent difficile de réduire dans un calcul ce qui est dans la figure" ou encore "Je croy qu'on pourroit manier par ce moyen la mécanique presque comme la geometrie, et qu'on pourroit mesme venir jusqu'a examiner les qualités des matérieux, par ce que cela dépend ordinairement de certaines figures, de leurs parties sensibles. Enfin je n'espere pas qu'on puisse aller assez loin en physique, avant que d'avoir trouvé un tel abrégé pour soulager l'imagination. Car nous voyons par exemple quelle suite de raisonnements géométriques necessaire pour expliquer seulement l'arc en ciel, qui est un des plus simples effects de la nature, par où nous pouvons juger combien de consequences seroient nécessaires pour penetrer dans l'interieur des mixtes, dont la composition est si subtile que le microscope, qui en decouvre bien plus que la cent-millieme partie, ne l'explique pas encor assés pour nous aider beaucoup. Cependant il y a quelque esperance d'y arriver en partie, quand cette analyse veritablement géometrique sera établie."

Naturellement, ainsi séparés, algèbre et géométrie entretiennent de curieux rapports de prédation réciproque : l'algèbre aboutit toujours en évitant les pièges des figures et la géométrie, en retour, garantit le caractère "réel", tangible des calculs algébriques qui, sans elle, ne seraient que de pures abstractions. La géométrie fait qu'"ils sautent aux yeux". Les objets "réels" de notre espace et leurs rapports de distance incarnent une <u>évidence</u> qui force la conviction et restent par là-même les <u>dépositaires</u> de la vérité mathématique. L'existence "réelle" de la dynamique, de la cinématique étaient censée bien montrer que les formes algébriques n'étaient pas de "pures vues de l'esprit" et réglaient les lois de la nature.

Avec le plan complexe d'Argand-Wessel, puis avec les quaternions de Hamilton et l'algèbre de Grassmann, une bouleversement se produit, on "y voit" enfin la multiplication des imaginaires (qui deviennent alors prétendument imaginaires). La relation $i^2 = -1$ n'apparaît plus comme une relation satisfaite par une "grandeur impossible" [3]. La multiplication des grandeurs n'était auparavant saisie qu'au travers de changement d'unité. Désormais, un <u>geste</u> géométrique - une rotation ! - accompagne le mystérieux symbole i

i² = -1   s'écrit

   L'intuition géométrique n'est plus seulement l'espèce d'inspection oculaire qui vérifiait l'accord des calculs mathématiques et des mesures d'objets réels, elle est désormais bien plus la possibilité de saisir par diagramme les gestes de conquête de l'espace et consacre ainsi l'émergence d'un continuum physico-géométrique à deux dimensions aussi éloigné du plan d'Euclide que de "l'espace vectoriel abstrait ℝ × ℝ " de nos formalistes !

   Les "Lectures on Quaternions" réussissent ensuite à donner à ces actes d'investigations de l'espace le caractère systématique de l'algèbre. Partant des points - le geste primordial de prise de position - il définit les vecteurs et leur addition. On savait naturellement depuis longtemps construire des parallélogrammes de forces et de vitesses[4] mais les vecteurs n'étaient compris que comme suivant la virulence d'une force ou la fulguration d'une célérité et non comme notion géométrique autonome.

   Notre espace affine naît avec Hamilton. L'homothétie, l'addition et la multiplication vectorielles naissent par une sorte d'ascèse qui révèle un geste certes dépouillé, le "vection act", le vecteur comme visée virtuelle et transport, mais nullement "abstrait" de la réalité sensible. Le continuum physico-géométrique est bien apprécié comme domaine d'exercice d'une véritable géométrie expérimentale[5].

   Hamilton introduit la notion de vecteurs en considérant deux points A et B. A est la position donnée de l'observateur, "l'analyseur" et B ce que je vise, ce vers quoi je tends, ce que je détermine "l'analysand". B - A est le symbole par lequel je m'approprie B par la pensée en le joignant à A par une ligne droite. L'addition, et la formule A + (B - A) = B, s'écrit A + $\vec{V}$ = B (avec $\vec{V}$ = $\overrightarrow{AB}$). B est alors produit. Avec le signe -, B était simplement observé de A. Il est maintenant conquis par vection et ce transport s'écrit A + $\vec{V}$ = B.

   "In this way, the symbol B-A has come with us to denote the straight line from A to B ; the point A being (at first) considered as a known thing, or a datum in some geometrical investigation, and the point B being (by contrast) regarded as a sought thing, or a quaesitum : while B - A is at first supposed to be a representation of the ordinal relation in space, of the sought point B to the given point A ; or of the geometrical DIFFERENCE of those two points, that is to say, the difference of their TWO POSITIONS in space ; and this difference is supposed to be exhibited or constructed by a straight line. Thus, in the

astronomical example of earth and sun, the line B-A has been seen to
extend from the place of observation A (the earth), to the place of the
observed body B (the sun) ; and to serve to CONNECT, at least in
thought, the latter position with the former.

Again you have seen that with me the primary geometrical operation
denoted by the mark +, and called by the name ADDITION, or more fully,
symbolical Addition, consists in a certain correspondent ordinal
SYNTHESIS of the position of a mathematical point in space. Instead of
comparing such a position, B, with another position A, we now regard
ourselves as deriving the one position from the other. The point B had
been before a punctum analyzandum ; it is now a punctum constructum. It
was lately the subject of an analysis ; it is now the result of a
synthesis. It was a mark to be aimed at ; it is now the end of a flight,
or of a journey. It was a thing to be investigated (analytically) by our
studying or examining its position ; it is now a thing which has been
produced by our operating (synthetically) on another point A, with the
aid of a certain instrument, namely, the straight line B-A, regarded
now as a VECTOR, or carrying path, as is expressed by the employment of
the SIGN OF VECTION, +, through the general and identical formula :

$$(B - A) + A = B.$$

That other point A, instead of being now a punctum analyzans, comes to
be considered and spoken of as a punctum vehendum, or more briefly, and
with phrases of a slightly less foreign form, it was an analyzer, but is
now a VEHEND ; while the point B, which had been an analyzand, has come
to be called a VECTUM, according to the general formula :

$$Vector + Vehend = Vectum ;$$

where Plus is (as above remarked) the Sign of Vection, or the
characteristic of ordinal synthesis. From serving, in the astronomical
example, as a post of observation, the earth, A, comes to be thought of
as the commencement of a transition, B - A, which while thus beginning at
the earth is conceived to terminate at the sun ; and conversely the sun,
B, is thought of as occupying a situation in space, which is not now
proposed to be studied by observation, but is rather conceived as one
which has been reached, or arrived at, by a journey, transition, or
transport of some moveable point or body from the earth, along the
geocentric vector of the sun."

Hamilton définissait son algèbre comme le science du temps pur[6].
Il entendait par là que le temps du géomètre était scandé par les
différentes étapes d'une conquête progressive de l'espace. Pour Hamilton,
il ne suffit pas de comparer les autres points à un point de référence
(ce que permettent la définition des vecteurs $\overrightarrow{BA}$, de l'addition et de la

multiplication par des scalaires[7]). Il faut aussi explorer les entours
de A et introduire les rotations. Comme dans le cas des complexes, il
définit le rapport de deux vecteurs non colinéaires (figure ci-dessous).

"The Analysis and Synthesis, hitherto considered by us, have been of
an ORDINAL kind ; but we now proceed to the consideration of a different
and a more complex sort of analysis and synthesis, which may, by contrast
and analogy, be called CARDINAL. As we before (analytically) compared a
POINT B, with a point A, with a view to discover the ordinal relation in
space of the one point to the other ; so we shall now go on to compare
one directed line, or vector, or RAY, β, with another ray, α, to
discover what (in virtue of the contrast and analogy just now referred
to) I shall venture to call the cardinal relation of the one ray to the
other, namely, (as will soon be more clearly seen), a certain complex
relation of length and of direction."

Il est bien connu qu'un calcul compliqué est nécessaire pour
connaître les coefficients d'une rotation de l'espace déterminée par deux
vecteurs, alors que la géométrie la voit "d'un coup d'oeil". Grâce aux
quaternions, l'algèbre est aussi prompte que l'intuition géométrique.

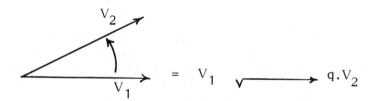

L'algèbre est bien la science de l'ordre en train de se construire.
L'espace n'est plus saisi seulement de l'extérieur par le quadrillage
cartésien ; les actes de choix d'échelle, de composition de transport
(addition vectorielle) et de multiplication de directions sont définis.
Il s'établit ainsi une articulation interne entre opération et intuition
qui confère à l'espace quaternionique une dynamique tout à fait
spécifique. Les vecteurs $\vec{i}$, $\vec{j}$, $\vec{k}$ ne sont plus simplement des vecteurs
de bases mais dirigent des rotations virtuelles[8] comme le montre leur
loi de composition

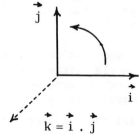

$$\vec{k} = \vec{i} \cdot \vec{j}$$

Hamilton pressentait le caractère physico-mathématique de sa découverte et J. C. Maxwell lui rend souvent hommage[9]. Les quaternions ne sont pas seulement le premier corps non commutatif :

"Now Quaternions, or the doctrine of Vectors, is a mathematical method, but it is a method of thinking, and not, at least for the present generation, a method of saving thought... It calls upon us at every step to form a mental image of the geometrical features represented by the symbols, so that in studying geometry by this method we have our minds engaged with geometrical ideas, and are not permitted to fancy ourselves geometers when we are only arithmeticians."

ou encore :

"A most important distinction was drawn by Hamilton when he divided the quantities with which he had to do into scalar quantities... and vectors.... The invention of the calculus of Quaternions is a step towards the knowledge of quantities related to space which can only be compared for its importance, with the invention of triple coordinates by Descartes".

Dans la lignée d'Ampère et de Faraday, les équations de Maxwell consacrent l'émergence d'un espace électro-géométrique[10].

"The only experimental fact which we have made use of in this investigation is the fact established by Ampère that the action of a closed circuit on any portion of another circuit is perpendicular to the direction of the latter. Every other part of the investigation depends on purely mathematical considerations depending on the properties of lines in space. The reasoning therefore may be presented in a much more condensed and appropriate form by the use of the ideas of language of the mathematical method specially adapted to the expression of such geometrical relations - the Quaternions of Hamilton." (J. C. Maxwell)

Pour produire ses concepts, le physicien provoque en quelque sorte le champ par différentes espèces d'interventions spatiales. Cette géométrie expérimentale ne se manifeste pas tant par des prises immédiates de position mais par des saisies différentielles ! translations infinitésimales pour le gradient, petites boucles entourant un circuit pour le rotationnel, volumes infinitésimaux qui enveloppent une source pour définir la divergence[11].

La théorie du champ de Maxwell rend manifeste la communauté profonde et nécessaire entre géométrie et physique. Une classification mathématique des grandeurs physiques s'impose. C'est l'objet du célèbre article "Sur la classification mathématique des quantités physiques" :

"According to Ampère and all his followers, however, <u>electric currents are regarded as a species of translation</u>, and, <u>magnetic force as depending on rotation</u>. I am constrained to agree with this view, because the electric current is associated with <u>electrolysis, and other undoubted instances of translation</u>, while <u>magnetism is associated with the rotation of the plane of polarization of light</u>, which, as Thomson has shewn, involves actual motion of rotation.

The Hamiltonian operator ∇, applied to any vector function, <u>converts it from translation to rotation, or from rotation to translation, according to the kind of vector to which it is applied</u>.

I shall conclude by proposing for the consideration of mathematicians certain phrases to express the results of the Hamiltonian operator ∇. I should be greatly obliged to anyone who can give me suggestions on this subject, as I feel that the onomastic power is very faint in me, and that it can be successfully exercised only in societies.

∇ is the operation $i \dfrac{d}{dx} + j \dfrac{d}{dy} + k \dfrac{d}{dz}$, where i, j, k are unit vectors parallel to x, y, z respectively. The result of performing this operation twice on any subject is the well known operation (of Laplace)

$$\nabla^2 = -\left(\frac{d^2}{dx^2} + \frac{d^2}{dy^2} + \frac{d^2}{dz^2} \right).$$

The discovery of the square root of this operation is due to Hamilton.

And, first, I propose to call the result of ∇^2 (Laplace's operation with the negative sign) the <u>Concentration</u> of the quantity to which it is applied.

For if Q be a quantity, either scalar or vector, which is a function of the position of a point ; and if we find the integral of Q taken throughout the volume of a sphere whose radius is r ; then if we divide this by the volume of the sphere, we shall obtain $\bar{Q}$, the <u>mean</u> value of Q within the sphere. If Q_o is the value of Q at the centre of the sphere, then, when r is small,

$$Q_o - \bar{Q} = \frac{r^2}{10} \nabla^2 Q,$$

or the value of Q at the centre of the sphere exceeds the mean value of

Q within the sphere by a quantity depending on the radius, and on $\nabla^2 Q$. Since, therefore, $\nabla^2 Q$ indicates the excess of the value of Q at the centre above its mean value in the sphere, I shall call it the concentration of Q.

If Q is a scalar quantity, its concentration is, of course, also scalar. Thus, if Q is an electric potential, $\nabla^2 Q$ is the density of the matter which produces the potential.

If Q is a vector quantity, then both Q_0 and $\bar{Q}$ are vectors, and $\nabla^2 Q$ is also a vector, indicating the excess of the uniform force Q_0 applied to the whole substance of the sphere above the resultant of the actual force Q acting on all the parts of the sphere.

Let us next consider the Hamiltonian operator ∇. First apply it to a scalar function P. The quantity ∇P is a vector, indicating the direction in which P decreases most rapidly, and measuring the rate of that decrease. I venture, with much diffidence, to call this the <u>slope</u> of P. Lamé calls the <u>magnitude</u> of ∇P the "first differential parameter" of P, but its <u>direction</u> does not enter into his conception. We require a vector word, which shall indicate both direction and magnitude, and one not already employed in another mathematical sense. I have taken the liberty of extending the ordinary sense of the word slope from topography, where only two independent variables are used, to space of three dimensions.

If σ represents a vector function, $\nabla\sigma$ may contain both a scalar and a vector part, which may be written $S\nabla\sigma$ and $V\nabla\sigma$.

I propose to call the scalar part the <u>Convergence</u> of σ, because, if a closed surface be described about any point, the surface integral of σ, which expresses the effect of the vector σ considered as an inward flux through the surface, is equal to the volume integral of $S\nabla\sigma$ throughout the enclosed space. I think, therefore, that the convergence of a vector function is a very good name for the effect of that vector function in carrying its subject inwards towards a point.

But $\nabla\sigma$ has, in general, also a vector portion, and I propose, but with great diffidence, to call this vector the <u>Curl</u> or <u>Version</u> of the original vector function.

It represents the direction and magnitude of the rotation of the subject matter carried by the vector σ. I have sought for a word which shall neither, like Rotation, Whirl, or Twirl, connote motion, nor like Twist, indicate a helical or screw structure which is not of the nature of a vector at all.

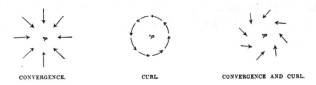

CONVERGENCE.   CURL.   CONVERGENCE AND CURL.

If we subtract from the general value of the vector function σ its value σ_0 at the point P, then the remaining vector $\sigma - \sigma_0$ will, when there is pure convergence, point towards P. When there is pure curl, it will point tangentially round P ; and when there is both convergence and curl, it will point in a spiral manner.

The following statements are true:-

The slope of a scalar function has no curl.

The curl of a vector function has no convergence.

The convergence of the slope of a scalar function is its concentration

The concentration of a vector function is the slope of its convergence, together with the curl of its curl."

La première moitié du XIXe siècle est donc le témoin d'un profond bouleversement de la géométrie : de Science des Figures (rapports figés de position) elle devient science de la capture physique de l'espace.

Dans un domaine, a priori très éloigné, la théorie de Galois effectue le même type de révolution. Sa théorie ne se propose plus de rechercher des "solutions" (écrire une expression du type x = ...) mais de décrire la dynamique même de cette recherche. Les quaternions, les équations de Maxwell capturent l'espace, Galois pense l'individuation progressive des racines en mathématicien.

Cette théorie des Equations (qu'il vaudrait mieux appeler théorie de l'apprentissage des équations comme on le verra plus tard) prend explicitement pour thème la symétrie et la dissymétrie de l'ensemble des racines d'une équation irréductible à coefficients entiers.

(E) $x^n + a_{n-1}x^{n-1} + \ldots + a_0 = 0.$

Comme nous allons le voir, la symétrie n'est pas pensée par Galois comme un vague sentiment de "satisfaction esthétique" ou une propriété "astucieuse" qui permet de "simplifier les problèmes". Elle est tout à fait reliée à l'idée de virtualité du précédent paragraphe. Rappelons que les travaux de Cauchy avaient déjà montré l'existence d'un domaine $D(\alpha_1, \ldots, \alpha_n)$ étendant les rationnels et sur lequel (E) se décompose :

$$(x - \alpha_1)(x - \alpha_2) \ldots (x - \alpha_n) = 0.$$

L'indexation des racines sous la forme (α_i) est tout à fait arbitraire. Pour l'algébriste qui calcule sur les rationnels, ces racines n'existent pas. Ce qui existe c'est le domaine $D(\alpha_1, \ldots, \alpha_n)$ et la manière dont il s'obtient à partir de E. A proprement parler, ces racines ne sont pas "possibles". Elles créent du "possible" au sens où le domaine $D(\alpha_1, \ldots, \alpha_n)$ est d'autant plus vaste que les substitutions que peut effectuer un algébriste ne connaissant que les rationnels sont plus nombreuses. Ces substitutions peuvent être toutes les permutations qui échangent les racines. Il peut aussi exister des relations rationnelles "particulières" entre elles, (équations "bicarrées", équations "réciproques"). Le groupe de symétrie de l'équation (groupe de Galois) est alors le plus grand groupe de substitutions qui respecte les relations entre les racines. Il traduit notre manque de discernement entre les (α_i) mais apprécie également la dimension du nouveau domaine de rationalité $D(\alpha_1, \ldots, \alpha_n)$ qu'il constitue. Il mesure bien l'espace de liberté engendré par le problème. Ces racines n'existent que virtuellement. Elles n'agissent pas comme individus mais par la potentialité de leurs échanges.

Ces racines seront "réelles", elles passeront dans l'existence concrète dans le domaine $D(\alpha_1, \ldots, \alpha_n)$ lorsque je ne serai pas seulement capable d'écrire un signe "formel" mais aussi d'exhiber un procédé de discernement entre ces racines. Ce procès de discernement est bien une cassure de la symétrie du groupe des racines. Discerner plus, c'est se montrer capable d'exhiber certaines "grandeurs tests", non rationnelles, et invariantes par un groupe de symétrie plus restreint. Les racines vont s'individuer au fur et à mesure de la précision des "expressions algébriques test" construites par certaines adjonctions aux rationnels.

Prenons l'exemple d'une équation très particulière : l'équation bicarrée $x^4 + 2px^2 + q = 0$. Si x_1, x_2, x_3, x_4 constituent une indexation arbitraire des racines, je puis découper deux parties $\sigma_1 = (x_1, x_2)$, $\sigma_2 = (x_3, x_4)$ telles que $x_1 + x_2 = 0$, $x_3 + x_4 = 0$.

La distinction entre σ_1 et σ_2 est complètement "arbitraire" (l'indexation de ces couples n'est pas imposée par l'équation donnée). On peut montrer facilement que le groupe de symétrie est à huit éléments. Pour différencier "concrètement" entre σ_1 et σ_2 j'adjoins la quantité $\sqrt{p^2-q}$ aux rationnels. Je donne ainsi un sens à "l'expression test" :

$$x_1^2 - x_3^2 = \sqrt{p^2-q} = 2\sqrt{\Delta}.$$

Cette fois-ci σ_1 et σ_2 ne jouent plus "le même rôle" dans le nouveau domaine. L'algébriste qui calcule dans ce domaine pourra indexer canoniquement les couples σ_1 et σ_2. Un groupe de symétrie à quatre éléments est alors permis (échange de x_1 et x_2, x_3 et x_4 qui jouent

encore   respectivement   "le même rôle).   En introduisant la quantité
$\sqrt{-p-\sqrt{\Delta}}$, je donne un sens à l'expression $x_3 - x_4 = 2\sqrt{-p-\sqrt{\Delta}}$ brisant la
symétrie entre x_3 et x_4 . Les racines x_3 et x_4 sont donc
maintenant proprement déterminées.

    Résoudre une équation, ce n'est plus poser un signe = arbitraire
entre une "inconnue" et une "valeur", c'est exhiber une séquence de
groupes décroissants $G_k \supset G_{k-1} \supset G_o = \{1\}$ obtenus grâce à une séquence
de fonctions-test f_k , ..., f_1 telle que f_i soit invariante par le
groupe G_i et non par G_{i+1} . Nous pouvons expliciter la résolution de
l'équation bicarrée sous forme de diagramme :

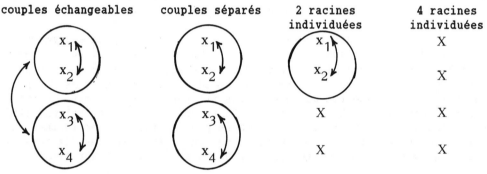

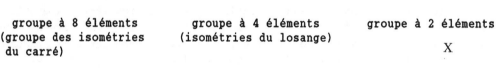

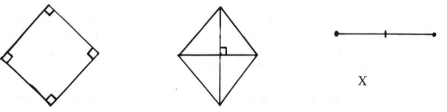

    La déconnection progressive du diagramme qui interdit les gestes
d'échange correspond à l'individuation d'une figure. Ces diagrammes
correspondent à une séquence de "résolvantes". Nous aurions pu partir
d'un tétraèdre régulier.

Résoudre une équation c'est désormais étudier <u>l'apprendre de la résolution des équations</u>, c'est-à-dire exhiber une séquence de groupes décroissants qui doivent laisser des mondes de racines de plus en plus précis.

Ce que Galois avait mis en évidence pour les équations algébriques, P. Curie l'énoncera pour la Physique : "C'est la dissymétrie qui crée le phénomène". "Il existe une symétrie maximale compatible à un phénomène" :

- Un champ électrique est incompatible avec l'existence d'un centre de symétrie et un plan de symétrie perpendiculaire à l'axe du champ.

- Un champ magnétique est incompatible avec un plan passant par l'axe du champ.

La dissymétrie n'est plus appréciée comme un manque, une case vide dans un programme de classification. Elle opère pour elle-même. <u>Symétrie et dissymétrie doivent se comprendre comme un double mouvement de déploiement et d'actualisation.</u> De l'extérieur, la symétrie apparaît comme un crible a priori de possibles. De l'intérieur, elle manifeste la potentialité (à dévoiler !) d'un monde de phénomènes ou d'une famille de solutions. Unifier des forces, résoudre une équation, c'est conformément au principe de Raison Suffisante, faire apparaître au cours du dévoilement la séquence des expériences ou des quantités algébriques qui assurera <u>la maîtrise progressive de cette potentialité</u> en produisant une séquence de mondes dont les êtres sont de plus en plus déterminés.

<u>CONCLUSION</u>

Depuis quelques années, la disgrâce dont semblait victime l'intuition géométrique semble s'estomper. On pressent que le geste géométrique est celui de la constitution même de l'espace et ne se réduit jamais à la simple "illustration" d'un raisonnement. L'enseignement classique de la géométrie se bornait surtout à étudier les rapports de distance d'une extension déjà fournie. Les diagrammes du complexe physico-géométrique explorent le continuum avant toute prise extérieure de position et plus exactement, leurs tâtonnements, leurs investigations, ne découvrent pas des saillies "cachées", elles sont ces singularités[12].

Il existe une puissance opératoire tout à fait particulière aux diagrammes [13]. Ils ne se contentent pas de visualiser des algorithmes ou de coder et de compactifier "l'information" pour la restituer sous forme de modèles ou de "paradigmes"[14]. Le diagramme est bien ce grouillement de gestes virtuels : pointer, boucler, prolonger, strier le continuum. Une simple accolade, un bout de flèche et le diagramme saute par-dessus les figures et contraint à créer de nouveaux individus[15].

Le diagramme ignore superbement toutes les vieilles oppositions "abstrait-concret", "local-global", "réel-possible". Il garde en réserve toute la plénitude et tous les secrets des fonds et des horizons que sa magie tient toujours pourtant en éveil[16].

Peut-être retrouvons-nous ainsi la symétrie des géomètres grecs. Elle ne se dépréciait pas en vague confort du regard, en satisfaction "esthétique". Elle n'épiçait pas le Monde par une "touche d'élégance". Elle manifestait simplement que quelque chose prétend à l'existence dans l'Harmonie.

NOTES

(1)   Einstein ne fut jamais satisfait de sa célèbre équation reliant le tenseur de courbure au tenseur d'énergie-impulsion. Il la comparait à un temple à deux ailes : l'aile de bois (la matière) n'était pas en harmonie avec l'aile de marbre (la géométrie).

(2)   Les imaginaires étaient connus au XVIe siècle. Les premiers traités de Wessel et d'Argand datent de 1797 et 1806.
        Dans son livre "Spinors and Space-time", R. Penrose rend d'ailleurs hommage à ces précurseurs. Comprendre notre espace physique ne consiste nullement à écrire n = 4 dans une formule générale mais à repenser physiquement le "formalisme" des espaces vectoriels et des fibrés, à ne plus considérer l'espace comme un "cadre abstrait" mais comme un champ d'expériences cinématiques, optiques, électriques, etc.

(3)   Grandeur résultant de la mesure physique de dimensions d'objets existant dans la nature et dérivant d'une échelle de comparaison : le "scalaire réel".

(4)   Connus respectivement depuis Archimède et les cinématiciens du Moyen-Age.

(5)   cf. l'écriture quaternionique des travaux d'Ahyah sur les solitons.

(6)   cf. préface des Lectures on Quaternions "it early appeared to me
        that these ends might be attained by our consenting to regard
        algebras as being no mere art, or language, nor primarily a science
        of quantity but rather as science of order in progression".
        renvoie aussi à l'esthétique de Kant.

(7)   Scalaire : de scalae, échelle.

(8)   Opérateurs infinitésimaux.

(9)   Il y consacra les vingt dernières années de sa vie. Il y voyait la
        possibilité de contempler les quantités physiques et géométriques
        d'une manière plus originaire : "But for many purposes of physical
        reasoning, as distinguised from calculation, it is desirable to
        avoid explicitly introducing the Cartesian coordinates, and to fix
        the mind at once on a point of space instead of its three
        coordinates, and on the magnitude and direction of a force instead
        of its three components. This mode of contemplating geometrical and
        physical quantities is more primitive and more natural than the
        other, although the ideas connected with it did not receive their
        full development till Hamilton made the next great step in dealing
        with space, by the invention of his Calculus of Quaternions.
                As the methods of Descartes are still the most familiar to
        students of science, and as they are really the most useful for
        purposes of calculation, we shall express all our results in the
        Cartesian form.   I am convinced, however, that the introduction of
        ideas,   as   distinguised   from   the   operations and methods of
        Quaternions, will be of great use to us in the study of all parts
        of our subject, and especially in electrodynamics, where we have to
        deal with a number of physical quantities, the relations of which
        to each other can be expressed far more simply by a few expressions
        of Hamilton's, than the ordinary equations". (J. C. Maxwell)

(10)   cf. l'éloge particulièrement vibrant de Maxwell à Faraday qui a vu
        des champs et des forces là où d'autres n'avaient vu que de la
        distance :

        "From the straight line of Euclid to the lines of force of Faraday
        this has been the character of the ideas by which science has been
        advanced, and by the free use of dynamical as well as geometrical
        ideas we may hope for a further advance."
                "The geometry of position is an example of a
        mathematical science established without the aid of single
        calculation. Now Faraday's lines of force occupy the same position
        in electro-magnetic science that pencils of lines do in the

geometry of position. They furnish a method of building up an exact mental image of the thing we are reasoning about. The way in which Faraday made use of his idea of lines of force in co-ordinating the phenoma of magneto-electric induction shews him to have been in reality a mathematician of a very high order - one from whom the mathematicians of the future may derive valuable and fertile methods".

(11) $\overrightarrow{\text{grad}}\ V = \vec{E}$ $\overrightarrow{\text{rot}}\ \vec{B} = \vec{j}$ $\text{div}\ \vec{E} = \rho$

(12)   à la fois geste et objet visé

(13)   cf. "Black holes - The membrane paradigm" jusqu'aux travaux de Penrose, le point de vue de l'étoile gelée (frozen star - collapsed star) donnait la théorie des trous noirs. Une modification du diagramme "frozen star" a fait surgir la dynamique dans la théorie. Les calculs sont venus ensuite....

(14)   C'est peut-être un complément de l'ordinateur qui excelle à traiter les algorithmes les plus complexes, éprouve des difficultés à imiter les gestes quotidiens (la main qui cherche... reconnaître une forme...). Constituer un monde suppose toujours un horizon de formes non déjà reconnues mais virtuellement accessibles. Cette notion : la "profondeur" semble très récalcitrante aux codes informatiques.....

(15)   Cf. Feynmann : "l'électron fait ce qu'il veut. Il va où il veut..".

(16)   On pourrait effectuer ce raccourci amusant de l'histoire de la notion de vecteur.

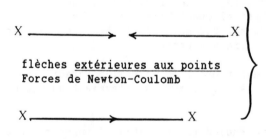

flèches extérieures aux points
Forces de Newton-Coulomb

le vecteur transport de Hamilton

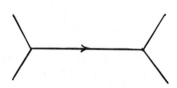

diagramme de Heisenberg-Yukawa pour les particules virtuelles

MATHEMATICS AS A SERVICE SUBJECT: SOME OPINION PAPERS GIVEN BY MEMBERS OF THE FRENCH ACADEMY OF SCIENCES

presented by

G. Choquet
French Academy of Sciences, Paris, France

INTRODUCTION

As part of the study undertaken by the CIEM on the theme, "Mathematics as Service Discipline", the Academy of Sciences, in concert with J.P. Kahane, chairman of CIEM, and P. Germain, permanent secretary of the Academy, asked several of its members for opinion papers on the role played by mathematics in their research and for an evaluation of how mathematicians may help them. Eight researchers from the disciplines of Physics, Mechanics, and Cellular and Molecular Biology prepared papers : A.Abragam, A. Blanc-Lapierre, C. Cohen-Tannoudji, J. Friedel, P. Germain, A. Maréchal, L. Michel, and B. Pullman.

From these scientists' accounts, although as varied as their respective fields of research, several general conclusions can be drawn that will doubtless be of great interest to mathematics instructors and lead them to reconsider their teaching methods and to establish closer co-operation with their colleagues from experimental disciplines.

One of the principal ideas lies in the importance of a solid grounding in mathematics, especially in the fundamentally important field of geometry. An intuitive grasp of geometry, a notion of space, and a direct vision of objects are indispensable for both experimental and theoretical physicists.

A solid knowledge of classic analytical methods and a familiarity with calculus seem indispensable for all researchers, who should also be well acquainted with algebraic elements and linear algebra.

Once within the different disciplines, the mathematical skills needed quickly diversify : linear representations of groups, algebraic topology, differential varieties, modular functions, equations with partial derivatives, quasi-periodicity, finite-element methods, fractal objects, attractors, and, in short, many other of the most important and most discussed disciplines to be found in the pages of Math. Reviews.

A preliminary remark can thus be made about the formal mathematical training received by physicists in universities and *grandes écoles* : much of it leaves them without the necessary tools to become acquainted with the latest developments of mathematical inquiry. This point should be remembered, for it calls into question the teaching dispensed at universities and *grandes écoles*. Indeed, in reading between the lines of these papers we can distinguish a constant criticism, at times explicit, of the way mathematics is taught at a post-secondary level.

Of the various mathematical tools used by physicists, the lion's share is taken up by Probabilities and Fourier Analysis. Probability is touched on in secondary schooling, yet, paradoxically, a student in undergraduate mathematics can complete his training without a basic course in probabilities. Fourier Analysis owes its importance to the wave nature of such phenomena as electro-magnetism and to its close relation to convolution, which makes it an increasingly valuable tool both in theoretical work and in very practical applications (e.g. tomography, imaging by magnetic resonance).

Curiously enough, no report speaks of the importance of computer science for experimental researchers. This can be put down to the fact that the computer, like the electron microscope or the particle accelerator, is not so much a goal of research for the scientist as a common everyday tool.

From general criticisms to specific recommendations about the teaching of mathematics : the scientists questioned outlined several broad areas of agreement, although individual differences cropped up in each paper.

- It is more important to know the axioms and the nature of theorems, as well their validating conditions, than to know their proofs. Crucial counter-examples should be given, various traps pointed out, and blind alleys exposed.

- Undue emphasis on rigor, formalism, and abstraction should be avoided, as it often stifles imagination and creativity. Advances are often based on intuitive and analogical reasoning and a profound knowledge of simple models. A good mathematics teacher should induce motivation in his students' minds, clarify by simple and stimulating examples ; only afterwards should logical deduction be deployed.

- French mathematicians present mathematics as a well-elaborated whole, divorced from any real motivation.

Acerbic in his comments, P. Germain puts mathematicians in the dock : They have thought of their discipline as independent of external sources of inspiration and lived in a world unto themselves. The results of this attitude have been educationally disastrous. Secondary-school math teachers have been led to exclude mechanics and even, for a disastrous period of time, geometry. In universities mechanics and physics hold a reduced shared of mathematicians' attention. In their publications a hermetic style is the norm, incomprehensible to non-specialists and often lacking even a cursory heuristic introduction for a larger public.

In universities physicists are led--and prefer--to teach themselves the mathematics their students need. Mathematics is thus pushed further out of the mainstream--this is one of the reasons that some math users, feeling uncomfortable in the venerable SMF, have opted to create their own "Industrial and Applied Mathematics Association" (SMAI) and that more applied mathematics publications are appearing, written in a non-canonical but stimulating and accessible style.

Some of the content of our scientists' papers is valid for all countries ; some for France alone.

How to conclude ?

Everywhere in the world "the problems arising from the physics of fundamental interactions have given inspiration to mathematicians and encouraged a mutually-enriching contact between our two communities".

Everywhere in the world "co-operation between mathematician and physicist is both useful and pleasant ; yet a few cafeteria encounters are simply not enough to make it work" --co-operation requires, as the saying goes, "a true investment".

It is in France, however, that one can observe a wilful and conscious rejection of new theories, however nascent and promising, on grounds of their not being formalizable or suited to enter into already toorigid canonical structures. Despite our great predecessors and trail-blazers in modern fields of research, we in France have been slow in developing these very domains : probabilities, game theory, logic, and computer science.

It is in France, as well, that mathematicians in secondary and post-secondary teaching persist in putting up obstacles to physicists. It is fairly symptomatic that the Stokes Formula, though fundamental to both mathematics and physics, has practically disappeared from the university's differential calculus program because of the impossibility of giving, at the undergraduate level, a sufficiently general and rigorous exposition of it.

When will we see, in lycées, universities, and, later, in graduate and research facilities, consultation, even co-operation, between mathematicians and physicists ? Each points his finger at the other through a fog of mutual incomprehension. Too many math professors are satisfied with a limpidly rigorous set of lecture notes, unconcerned with how the material is absorbed and why. The university's compartmentalization of the teaching load only exacerbates the problem.

True, mentalities are changing : our young mathematicians believe they have freed themselves from the rigid yoke imposed on their elders. But cultural traditions are powerful and insidious : we are heirs of Descartes, which is all very well and good as long as we do not become slaves to the tools bequeathed to us by our ancestors and trapped in formal, elaborate theory. Changes in the sciences arise from new ideas, still uncodified and incomplete, that we must learn to recognize, protect, and develop. They come from people who combine intuition, a fresh way of looking at the world, a rejection of pre-conceived notions, and an ability to exercise great powers of concentration. Yet the human mind is frail--it needs suggestions offered by the external world, in particular, the experimental sciences.

It is this constant interchange back and forth between observation, experimentation, and theory that is the true motive force behind all sciences and all wisdom.

(tranlated from the French by Stephen O'Shea)

★ ★ ★

Témoignage de A. ABRAGAM

En ce qui concerne l'enseignement des Mathématiques aux physiciens beaucoup dépend du type de physicien auquel s'adresse cet enseignement. Certains sont aussi soucieux de rigueur que les mathématiciens, d'autres, même théoriciens, ont d'autres besoins. Saviez-vous que Heisenberg l'un des plus grands théoriciens de notre siècle ignorait tout du calcul des matrices et lorsque Max Born lui fit remarquer que le calcul des symboles non-commutatifs qu'il avait introduits était celui des matrices, Heisenberg répondit plaintivement "il va falloir que j'apprenne ces machins là aussi ! Mais quod licet Jovi.. et ce qu'à pu se permettre Heisenberg n'est pas recommandé à tout le monde.

Ce qui me paraît la panoplie minimale du physicien aujourd'hui (à part l'informatique, essentielle, mais dont, à part l'usage de la machine à traitement de texte sur laquelle je vous écris ceci, je suis aussi ignorant que l'immense majorité des physiciens de mon âge) c'est le maniement de certains outils cités dans un ordre arbitraire, sans doute pas celui où ils devraient être enseignés :
équations différentielles et aux dérivées partielles linéaires, idées qualitatives sur les non-linéaires (essentiellement savoir qu'elles peuvent réserver des surprises), maniement des fonctions analytiques et calcul des résidus, notions sur les séries asymptotiques, transformée de Fourier et de Laplace, calcul des matrices, équations intégrales, éléments de calcul tensoriel et de géométrie de Riemann, éléments de théorie des groupes, algèbres des groupes de Lie et représentations irréductibles des groupes finis et du groupe des rotations, éléments de calcul des variations, notions qualitatives de topologie, maniement des fonctions spéciales, Bessel etc., valeurs propres et fonctions propres. Je n'en sais pas plus moi-même. Sur la manière d'enseigner : il est plus important de connaître l'existence et la nature des théorèmes ainsi que leurs conditions de validité que leur démonstration que l'on oublie de toute façon.

En ce qui concerne la rigueur j'attache de l'importance à ce que j'appelle irrespectueusement "savoir pinailler à bon escient". Montrer par des contre-exemples bien choisis à quelles erreurs qualitatives

peut conduire le manque de rigueur : connexion de l'espace pour
l'utilisation du théorème de Stokes-Ampère, passages à la limite non
autorisés qui conduisent quand ils sont appliqués sans discernement
à des résultats faux ou à des paradoxes.

★ ★ ★

Témoignage de J. FRIEDEL

1°) Ma *formation* en mathématiques a été très "classique" : après l'Ecole
Polytechnique de 1944 à 1946, j'ai passé, sans préparation spéciale,
une "licence de maths" à la Sorbonne en 1948.

2°) *Mon travail de recherche personnel* comme *mon enseignement de 3ème
cycle* en Physique des Solides ont porté sur des développements de
mécanique quantique élémentaire appliquée aux atomes, molécules et
milieux denses (solides et liquides), mais aussi sur des applications
de la mécanique classique des milieux denses (mécanique des milieux
continus solides et liquides, mécanique des milieux cristallins), et sur
celles des équations de diffusion.
        a) A ce titre, les *équations différentielles* de base ont joué pour
moi un rôle essentiel :
                - résolutions pratiques, notamment discrétisation.
                - résolutions approchées : méthodes de variations, séries de
perturbations diagrammatiques ; utilisation des fonctions de Green et
leurs expressions en fractions continues ; intégration dans le plan
complexe et résidus.
                - rôle des conditions aux limites pour les états étendus
(problèmes de diffusion).
                - étude des singularités des solutions ("catastrophes").
        b) L'aspect complémentaire est *l'algèbre linéaire*, les matrices,
les tenseurs et la notion associée d'opérateur.
        c) Les aspects *géométriques* ont joué un rôle aussi essentiel :
                - spécialiste des cristaux, je me suis fortement intéressé
aux raisonnements dans l'espace réciproque pour l'analyse des structures
atomiques et électroniques (transformées de Fourier).
                - spécialiste des défauts localisés dans les cristaux, je
me suis intéressé aussi aux distributions dans l'espace réel des
atomes et des électrons, comme à la géométrie des défauts considérés
comme des singularités d'un paramètre d'ordre.

3°) *Le travail de recherche de mon secteur* (et notamment de mes
élèves et des membres de mon labo) a souligné depuis 10 ans cette
importance de la *géométrie* :
                - espaces courbes, introduits en physique par la relativité mais
utilisés dans de nombreux problèmes : cristaux à contraintes internes

inhomogènes d'origine magnétique ou thermique, milieux plastiques à distributions continues de dislocations, agrégats, amorphes...
- espaces de dimension supérieure à 3 (description des modulations incommensurables des phases cristallines, quasicristaux, thermodynamique de l'ordre magnétique...).
- espaces fractals, lois d'échelle, groupe de renormalisation, attracteurs : phases magnétiques, percolation, agrégation...
- topologie, notamment des défauts de paramètres d'ordre.
- localisation dans l'espace de particules par (faibles) diffusions multiples.
- Processus de friction solide, désordre gelé (verres de spin), chaos classique, turbulence, où l'aspect géométrique est souligné.

Conclusions.

Dans mon travail *personnel*, les notions acquises durant ma formation se sont révélées suffisantes, sauf en ce qui concerne l'algèbre linéaire, alors déficiente (cela a changé !).

L'enseignement géométrique, que ce soit l'étude analytique des surfaces (courbures, projections, lignes particulières) ou la vision dans l'espace acquise par la vieille descriptive, a été fondamental pour un "visuel" et un praticien comme moi. Ce poids d'une géométrie assez mathématisée mais encore souple et visuelle s'est accru et ses aspects diversifiés depuis 15 ans en physique de la matière condensée.

Ce sont les concepts de base et leurs applications simples et précises dans des cas concrets qui me paraissent surtout utiles. Des applications spécifiques mais qui demandent des développements plus complexes peuvent s'apprendre quand c'est nécessaire : par exemple les fonctions spéciales (Bessel, Legendre...), les applications de la théorie des groupes. Par contre il me semblerait utile d'introduire aux étudiants des notions de topologie, les concepts de singularités, de fractal, les fonctions non périodiques conduisant au chaos...

★ ★ ★

Témoignage de P. GERMAIN

1°) *Sur le rôle joué par les mathématiques dans mes recherches personnelles.*

Mes travaux relèvent de la mécanique théorique et mettent en oeuvre la formation en analyse que j'ai reçue à l'Ecole Normale.

Certains contiennent des résultats originaux (modestes) de

Mathématiques (sur le problème de Tricomi-existence et unicité ; sur
les solutions élémentaires des équations de type mixte). Dans certains
travaux de Mécanique, j'ai eu à établir quelques théorèmes sur des
questions particulières. Dans toutes mes recherches, j'ai eu à faire
appel aux Mathématiques.

Mais ce qui a été pour moi le plus important -et ce que je serais
tenté de considérer comme un apport très significatif des Mathématiques
aux autres disciplines- c'est la capacité qu'offrent les Mathématiques
de mettre à jour ou d'acquérir une "intelligence mathématique" des
phénomènes mécaniques et physiques, en élaborant les concepts
essentiels, en formulant les voies les plus appropriées pour attaquer
une question ou pour la faire progresser.

C'est certes un choix inconfortable - les mathématiciens sont
tentés de ne pas vous prendre au sérieux si on s'attache à "voir" et
à "montrer" plutôt qu'à généraliser et démontrer et certains physiciens et ingénieurs ont parfois de la peine à découvrir et à
apprécier la fécondité d'une recherche et d'une réflexion menées dans
cet esprit.

2°) *Ce que j'attends des mathématiciens.*

J'ai eu les meilleures relations amicales avec mes collègues et
amis mathématiciens de ma génération, mais très peu de relations
scientifiques ; les problèmes dont je pouvais leur parler ne les
"accrochaient" pas. La situation a heureusement bien changé. Je peux
avoir ces dernières années des échanges scientifiques avec des
mathématiciens nettement plus jeunes que moi.

Les mathématiciens et spécialement les mathématiciens français
se sont -à mes yeux- "enfermés" dans leur discipline, négligeant ou
même excluant ceux qui travaillaient à leurs frontières, avec l'idée
implicite ou explicite que les mathématiques avaient atteint un stade
d'évolution suffisant pour pouvoir vivre sur elles-mêmes. Les
professeurs du secondaire ont laissé exclure la mécanique de leur
programme (ils ont été sur le point d'exclure aussi la géométrie). Les
futurs enseignants de mathématiques ne reçoivent plus d'initiation
à la mécanique et à la physique. Ils vivent et éduquent leurs élèves
dans leur tour d'ivoire. [...]

Cette position a des conséquences. En premier lieu, sur le
langage. En dépit des progrès tous réels que j'ai signalés et dont je
me réjouis, les mathématiciens -et spécialement les mathématiciens
français- tendent à vouloir imposer leur langage. Je suis prêt à
utiliser le langage des mathématiciens si celui-ci peut-être et est
effectivement "chargé" de signification mécanique et physique au
même titre que celui des ingénieurs qu'il est candidat à remplacer. Or
la majorité des professeurs de mathématiques ne sont pas préparés à le
faire. Ceci explique la répugnance des disciplines à accepter que la
formation mathématique de leurs étudiants soit assurée par des

professeurs de mathématiques. J'aurais de multiples exemples montrant
combien peut être néfaste un cours de mathématiques donné par un
matnématicien, soit disant pour préparer les étudiants à suivre un
cours de mécanique ou de physique (un cours sur les équations aux
dérivées partielles, sur les transformations de Fourier et de Laplace,
sur les tenseurs...).

En second lieu sur *l'esprit* des mathématiciens et sur leur
appréciation de certaines questions de mécanique et de physique. Là
aussi, j'aurais de multiples exemples. Ne comptent bien souvent à leurs
yeux que les travaux de caractère mathématique. Ils éprouvent ou
citent avec une certaine condescendance les travaux qui ont précédé
les leurs. La notion de "couche limite" et de "perturbations singu-
lières" a été introduite par Prandtl en 1904 - les concepts et
méthodes *essentiels* sont dûs à des mécaniciens théoriciens de Caltech.
Tout ceci est trop souvent oublié par ceux qui ont habillé ou
généralisé. Les mathématiciens ont eu souvent et ont encore un peu
tendance à se prendre comme de nobles chasseurs qui considèrent les
mécaniciens et les ingénieurs travaillant en amont comme de braves
rabatteurs de gibier.

Tout ceci entraîne une certaine *marginalisation* des mathématiques
pures avec corrélativement un risque de "secession". La création
de la "Société de Mathématiques Appliquées et Industrielles" qui a
pris un départ en flèche aux dépens de la fameuse et ancienne
"Société Mathématique de France" (à laquelle je continue d'adhérer
par simple fidélité) en est un exemple.

En conclusion, j'apprécie tout l'intérêt de l'étude lancée par la
Commission Internationale pour l'enseignement mathématique
"Mathématiques comme discipline de service". En fait, le mot "service"
n'est pas bon. J'ai critiqué l'attitude des mathématiciens qui considèrent un peu la mécanique comme une discipline de service. Je ne
souhaite pas inverser les rôles.

Je reviendrai à ce que je disais au début : les mathématiques
ont-elles l'intention et l'ambition de donner aux chercheurs des
autres disciplines des clés irremplaçables pour élaborer une
intelligence mathématique du monde scientifique ? Je me réjouis
-je le signale une fois encore- d'une heureuse évolution à cet égard.
Si oui, il convient donc d'infléchir le langage et l'esprit dans
lesquels sont enseignées les mathématiques, d'infléchir les modalités
de l'enseignement (la présentation des notions, la manière d'apprécier
et de noter etc...), de donner aux futurs enseignants une initiation
aux disciplines voisines pour qu'ils puissent avoir une idée de ce
qu'est "acquérir une intelligence mathématique" des sciences
physiques et mécaniques ; de leur donner une juste conscience de leur
appartenance à une communauté scientifique devant couvrir continûment
un spectre de connaissances, de méthodes d'approche, de découvertes,
au sein de laquelle les relations doivent être empreintes non

seulement d'amitié, mais de compréhension et d'estime mutuelles même
si chacun, bien naturellement, est porté à mieux apprécier le champ
dans lequel il a choisi d'oeuvrer.

★ ★ ★

Témoignage de A. MARECHAL

Mon expérience personnelle de recherche dans le domaine de
l'optique instrumentale (formation des images, cohérence, filtrage,
détection optique des petites molécules gazeuses etc...) m'amène
à évoquer le rôle crucial que joue la *transformation de Fourier* (T.F.).
Il semble qu'il y ait à cela deux raisons indépendantes :

1°) La T.F. s'introduit tout naturellement comme outil mathématique
bien adapté à la *nature ondulatoire* de la lumière, citons en particulier
    - la représentation des phénomènes de diffraction
    - l'expression du degré de cohérence partielle
    - la relation entre la répartition spectrale énergétique et
l'interférogramme, qui est à la base de la spectrométrie interféren-
tielle, technique maintenant largement utilisée.
    - l'étude des "fonctions de transfert" qui caractérisent le
comportement d'un instrument.

2°) La T.F. s'introduit également pour traduire simplement les
mécanismes de convolution fréquemment rencontrés (images d'objets
étendus par exemple).
Les spécialistes de l'optique utilisent fréquemment la T.F., le plus
souvent en appliquant les relations mathématiquement élémentaires ;
la T.F. est souvent pour eux un fil directeur très précieux (nous
avons par exemple pu concevoir des détecteurs interférentiels de
petites molécules à l'état gazeux en exploitant le caractère périodique
du spectre d'absorption, ceci s'avère très utile pour la détection
de polluants atmosphériques).
    J'ajoute que dans le domaine de la R.M.N. médicale (ou plutôt
l'imagerie par résonance magnétique) la T.F. joue un rôle irremplaçable
pour démêler des informations et arriver à dresser des cartes de
répartition de protons dans les tissus cérébraux par exemple (conférence
de Goldman à l'Académie en 1985).
    Inutile de répéter ici que l'on ne doit pas abandonner l'enseigne-
ment de la géométrie, discipline très précieuse pour la formation des

ingénieurs, tout en continuant à accorder une large place aux
distributions, dont la puissance et l'efficacité sont reconnues par tous.

★ ★ ★

Témoignage de A. BLANC-LAPIERRE

1°) Au début de mes travaux de recherches en physique, j'ai porté mon
attention sur des questions relatives à l'amplification des courants
faibles. Cela mettait en jeu de l'électronique et de l'électromagnétis-
me et une certaine analyse de divers phénomènes de fluctuations
intervenant comme facteurs limitant la sensibilité des appareils.

Je disposais du bagage correspondant à l'agrégation de physique
avec, en mathématiques, les deux certificats de
        Calcul différentiel et intégral
  et de
        Mécanique rationnelle.

Je n'avais reçu aucun enseignement mathématique ni sur les
intégrales de Fourier déjà utilisées par les ingénieurs de façon fortement intuitive, ni sur le *Calcul des Probabilités*. Je pense qu'à
cette époque (1934-38), cela n'était pas raisonnable et qu'un licencié
en mathématiques aura déjà dû avoir reçu au moins un minimum d'enseignement sur ces deux points.

Conclusion : nécessité d'un dialogue suffisant entre les enseignants de mathématiques et les besoins simultanés des utilisateurs.

2°) J'ai beaucoup travaillé avec FORTET et on connait les formations
respectives qui étaient les nôtres. Tous les deux, nous avions un
ferme désir de travailler ensemble et nous avons fait l'effort nécessaire . Je dois dire qu'il a bien fallu une bonne année, au moins,
pour que notre collaboration soit vraiment établie : pour moi,
nécessité de connaître assez de probabilités ou de mathématiques
diverses, nécessité pour posséder un certain langage et, pour FORTET,
obligation d'avoir une connaissance suffisante du sens physique
sous-jacent.

Conclusion : la collaboration entre un mathématicien et un
physicien est très utile et très agréable, mais *il ne suffit pas de
quelques rencontres à la cafétéria* pour la rendre effective.

★ ★ ★

Témoignage de A. COHEN-TANNOUDJI

Il me paraît important en premier lieu de rappeler quelle a été ma formation en Mathématiques. Dans le secondaire, j'ai suivi les anciens programmes et j'ai fait en particulier beaucoup de géométrie. J'en garde un excellent souvenir. Cela m'a donné le goût des représentations graphiques et diagrammatiques et une certaine facilité pour "voir les phénomènes dans l'espace". En classe préparatoire, j'ai été formé au calcul et j'ai acquis une discipline stricte pour le choix des notations, le maniement des équations, la présentation des résultats. Enfin, à l'Ecole Normale, j'ai eu la chance d'avoir des professeurs comme Cartan et Schwartz et d'être initié par eux à des domaines comme la topologie, les espaces de Hilbert, les distributions... Il me semble que j'ai suivi le bon ordre : tout d'abord, des problèmes concrets et des méthodes de calcul ; ensuite, les notions plus générales et plus fondamentales, les grandes synthèses.

Bien sûr, les Mathématiques sont essentielles pour les recherches que je poursuis en Physique Atomique, sur les Interactions entre matière et rayonnement. Le point qui me paraît crucial est qu'un enseignement de Mathématiques destiné à d'autres disciplines ne doit pas mettre l'accent uniquement sur la démarche déductive, sur le formalisme, sur l'abstraction. Les découvertes importantes procèdent souvent d'une démarche différente, basée au début sur l'intuition, la recherche de modèles simples, d'analogies, à la limite même de "bricolage", avant d'être reformulées ensuite de manière plus satis-faisante. Il ne faut pas à mon avis qu'un excès de rigueur étouffe au début l'imagination et la fantaisie. Je suis d'ailleurs persuadé que la recherche en Mathématiques Pures procède elle aussi au début de cette manière. Il me semble qu'un enseignement de Mathématiques devrait, avant l'exposé rigoureux d'une question donnée, expliquer au début les motivations, l'idée générale et le fil directeur qui sont à l'origine de l'élaboration des notions qui sont ensuite exposées de manière logique.

★ ★ ★

AN INQUIRY AMONG THE INDUSTRIAL PARTNERS OF THE "ASSOCIATION BERNARD GREGORY"

presented by

J. Ezratty
Association Bernard Gregory, Paris, France

The main purpose of the French "Association Bernard Gregory" is helping young scientists in finding employment in industrial fields. The "International Commission on Mathematical Instruction" asked us to proceed to a survey for its International Conference in Udine, Italy, April 1987 on "Mathematics as a Service Subject" among our industrial partners.

Our first questions focused on the mathematic education for non mathematics scientists, working in industry :

- How far are they able to carry on calculus without help ?
- How are they ready to translate into formulas factual problems ?
- How are they able to understand and use the formulas and the mathematic concepts read in scientific and technical papers ?
- How ready are they to face the up-to-date mathematics used in their specific field ?

The second aspect of the inquiry concerned professional mathematicians in their firms : what do the companies expect from them ?

You will find below some companies answers :

- Pechiney
- Informatique Internationale
- Commissariat à l'Energie Atomique

- Electricité de France
- Science & Tech.
- Michelin

★ ★ ★

Réponse de Pechiney

Y. FARGES

"Quelle formation mathématique faut-il donner aux scientifiques de l'Industrie ?

Nous ne pensons pas qu'il soit possible de répondre en quelques lignes et nous nous bornerons à énumérer quelques idées.

1°) Capacité de mener, sans assistance, certains types de calcul.

Cela n'est plus essentiel : l'informatique scientifique prend le relais, mais il faut que l'ingénieur *comprenne* :
- les concepts, les hypothèses, les méthodes utilisées ;
- *la signification des résultats* (l'ordinateur peut toujours faire le calcul et obtenir un résultat mais ce résultat peut n'avoir aucun sens). L'équilibrage d'un bilan matière d'un procédé peut être juste (du point de vue calcul) mais absurde (une mesure d'entrée, par exemple, étant fausse). Il faut savoir *détecter les erreurs*.

2°) Capacité de modéliser des situations concrètes. Cela est important.

Pour atteindre cette capacité, nous pensons qu'il faut étayer l'enseignement des mathématiques par des exemples réels. Les Américains exagèrent parfois dans ce domaine ; on optimise une chaîne de hamburgers, une implantation de stations de contrôle d'incendie ou le nombre de caisses sur un péage puis on présente, en annexe, les coefficients de Lagrange ! En France, on exagère parfois en sens inverse, en ne montrant pas à l'élève comment utiliser les concepts, comment construire les modèles, comment résoudre des *cas réels*. Même en mathématiques, dites appliquées, combien de cours, de livres de statistiques, de recherche opérationnelle,... où les exemples sont abstraits !

3°) Aptitude à approfondir les formules et les concepts ... à saisir les concepts mathématiques.

Certes, il faut sauvegarder cette *faculté d'abstraction* qui fait la force des scientifiques français et cette *culture générale* de base qui leur permet de se recycler, mais il faut aussi *adapter* le contenu au monde moderne. *L'analyse numérique* (méthode de Newton-Raphson, calculs aux différences finies,...) est une discipline, aujourd'hui, aussi vitale à l'ingénieur que le calcul infinitésimal, autrefois, avant l'apparition de l'informatique . N'en déduisons pas qu'il faut être exhaustif et tout enseigner. Il faut aussi chercher à écarter ou *à minimiser certains développements*.

4°) Qu'attendonc-nous des mathématiciens professionnels embauchés dans l'entreprise ?

Qu'ils puissent :
- définir, ou choisir, les méthodologies utilisées par les ingénieurs ;
- leur apporter les méthodes, les outils, en leur montrant les contraintes, les cas difficiles, les pièges ;
- résoudre, eux-mêmes, les problèmes mathématiques délicats où le scientifique de base n'a pas la compétence suffisante.

★ ★ ★

Réponse de Informatique Internationale

Y. DANDEU

CISI INGENIERIE est la filiale du Groupe CISI, spécialisée en Informatique Scientifique. Elle emploie 800 personnes, la plupart ayant suivi une formation en mathématiques (Grandes Ecoles, Université), une centaine d'entre eux travaillant plus particulièrement dans le domaine des mathématiques appliquées.

1°) Quelle formation mathématique faut-il donner aux scientifiques qui vont travailler dans l'industrie ?

Ainsi posée, la question est bien trop générale pour admettre une réponse un peu précise, cette dernière étant fonction du secteur industriel envisagé et à l'intérieur de ce dernier, du travail à effectuer.

Une fois éliminés les truismes (nécessaires et non suffisants) :
notions (pas trop vagues) d'algèbre linéaire, de calcul tensoriel,
d'espaces de Hilbert... et de tout ce qui concerne la calculasse
(théorique : intégrales, dérivées, Fourier and Co, Cauchy and Co...
ou numérique) pour ne pas parler de leur transcription via l'informatique, la seule réponse globale est, bien entendu, d'avoir appris
à apprendre ou, en reprenant les termes de votre lettre : avoir
développé l'aptitude à saisir, au cours d'une carrière les concepts
mathématiques (j'ajouterais l'algorithmique associée), qui émergeront
dans la discipline utilisée.

Pour les industries pour lesquelles nous travaillons (nucléaire,
spatial, défense, aéronautique), il est évident que la question n'est
pas "comment résoudre mathématiquement un problème bien posé" mais
successivement :
a) comment poser le problème (au sens des physiciens) de façon
à ce qu'il traduise "au moins mal" le phénomène à simuler.
b) l'énoncé ainsi obtenu étant supposé représentatif, comment le
traduire en "mathématiques convenables", pour que les problèmes
d'existence et d'unicité de la solution puissent prendre un sens.
c) comment choisir parmi toutes les traductions possibles, la
moins mal adaptée à la suite, autrement que par tatonnement.
d) finalement, comment construire localement une approximation
pas trop idiote de la solution, supposée exister et être unique.

Or d'après ce que j'ai pu entendre de nos collaborateurs ou consater
moi même,
a) n'est enseigné nulle part. Il est vrai que ça n'est pas facile
mais c'est pourtant là que le bât blesse le plus : l'essentiel des
difficultés rencontrées en pratique vient de ce que le problème est
physiquement mal posé.
b) n'est pas enseigné non plus sauf dans quelques cas résolus
pour lesquels la solution est bombardée. En général les élèves se
demandent bien pourquoi diable on est allé chercher des restrictions,
espaces,..., pareils.
c) et d) sont partiellement enseignés, au moins dans la plupart
des cours de type analyse numérique appliquée.
Néanmoins, il me semble d'expérience, qu'en moyenne, les élèves ne
comprennent pas ce qu'est une approximation et, en particulier, que
la plupart d'entre eux n'ont aucune notion des méthodes semi analytiques (construction et utilisation de développement asymptotiques).

2°) Qu'attendez-vous des mathématiciens embauchés dans l'entreprise ?

    - Qu'ils n'oublient pas trop vite ce qu'ils ont appris
    - Qu'ils aient appris à se documenter sans nurse
    - Qu'ils sachent faire ce pourquoi ils ont été embauchés et

donc qu'ils aient :
          . la formation pour le faire
          . l'honnêteté de dire qu'ils ne l'ont pas quand ils ne
l'ont pas (le problème de savoir s'ils l'ont presque et corollaire-
ment de la formation complémentaire nécessaire étant celui de leur
futur employeur).

★ ★ ★

Réponse du "Commissariat à l'Energie Atomique"

D. LAUNOIS

Pour remplir les missions qui lui sont confiées, aussi bien en
recherche appliquée que fondamentale, le CEA utilise les acquis et
participe au développement de nombreuses disciplines scientifiques,
notamment la physique, la chimie et la biologie.

Les mathématiques constituent bien évidemment un outil indispensable, mais nos travaux dans ce domaine relèvent bien plus des
mathématiques appliquées que des mathématiques pures et se ratta-
chent à ce que certains nomment la physique mathématique.

Les physiciens théoriciens dans leurs travaux (étude du chaos
quantique, grande unification,...) utilisent les derniers acquis des
mathématiques, mais, dans certains cas pointus, ils font appel à la
collaboration de "vrais" mathématiciens extérieurs au CEA.

Par ailleurs, pour les besoins du nucléaire civil ou militaire,
des chercheurs en mathématiques appliquées poursuivent deux objectifs
principaux :
     - la maîtrise d'algorithmes de calcul mis en oeuvre pour résoudre
       des systèmes complexes d'équations aux dérivées partielles qui
       décrivent des phénomènes physiques ;
     - l'exploration de nouvelles approches théoriques pour améliorer
       des modèles mathématiques.

A titre d'exemple, on peut citer quelques thèmes :
     - le passage d'une description microscopique au comportement à une
       échelle macroscopique (ce thème est dénommé par les mathéma-
       ticiens : homogénéisation) ;
     - la représentation des propriétés élastoplastiques anisotropi-
       ques de certains matériaux ;

- la représentation du comportement de matériaux à matrice tissée tridimensionnelle impliquant des modélisations à deux ou trois dimensions.

Enfin, un certain nombre d'études économiques, sur l'opinion, sur le traitement du signal,..., font appel aux derniers perfectionnements de l'analyse statistique.

Ce rapide survol des activités en mathématiques appliquées au CEA montre à l'évidence que la formation mathématique recherchée couvre les quatre points de la première question de la lettre en référence.

★ ★ ★

Extraits de la réponse de :
"ELECTRICITE DE FRANCE"
DIRECTION DES ETUDES ET RECHERCHES (DER)

La capacité des ingénieurs à dégager des idées générales, à les formaliser, à les traduire en modèles ou en schémas opérationnels, repose à peu près exclusivement sur leur formation mathématique.

La DER utilise très largement cette formation, car ses activités prolongent l'enseignement des écoles : beaucoup d'ingénierie, beaucoup de modélisation, peu d'investigation à caractère physique. [...]

L'avènement de la modélisation sur ordinateur renforcera le besoin d'une formation mathématique solide, pour deux raisons :
- parce que la construction des modèles dépend en majorité des compétences de l'ingénieur,
- parce que, si l'exploitation des modèles dans des études n'en dépend pas du tout, la capacité de synthèse intervient (et doit intervenir) comme antidote à ce que serait une exploitation "boîte noire" trop exclusivement expérimentale.

La DER a donc besoin d'ingénieurs, moins habiles dans l'art de résoudre les équations, mais plus habiles dans l'art de les écrire : de raisonner sur les modèles, de les critiquer, etc...

Dominer la pratique actuelle de l'ingénierie exige que l'on ait reçu un enseignement fortement structuré

Ce besoin ne peut qu'aller en s'accentuant. [...]

Un cours de mathématiques devant servir de support à une activité de modélisation numérique de phénomènes physiques, [...] pour avoir toute sa cohérence, sera associé à un cours d'analyse numérique. Il est clair que des notions de probabilités, statistiques font également partie du bagage de l'ingénieur. [...]

L'enseignement de la logique a été presque abandonné depuis 20 à 30 ans. Or c'est sur la logique que sont basés l'intelligence artificielle et les systèmes experts, donc :
- réintroduire un cours sur la logique et les systèmes formels orienté vers la modélisation. [...]
- développer aussi un cours sur la complexité des algorithmes et les problèmes NP-couples [...].

★ ★ ★

Réponse de SCIENCE & TEC.

A. ROUET

Je vous fais part de deux remarques que m'inspirent les questions posées par l'association sur les mathématiques et l'entreprise.

Ces remarques concernent en fait les mathématiques fondamentales, et non une forme de mathématiques appliquées souvent proche de l'informatique, déjà bien intégrée au milieu industriel.

Les mathématiques fondamentales sont un outil puissant dont l'efficacité en recherche industrielle m'est prouvée par les contrats obtenus par Science & Tec dans ce domaine :

Méthodes intégrales (applications en aérodynamique par exemple).

Topologie algébrique (applications en CAO : comment reconnaître l'intérieur de l'extérieur d'une structure en représentation "fil de fer").

Equations aux dérivées partielles.

Systèmes dynamiques (à la frontière de la physique théorique).

Etc...

Un industriel a rarement l'emploi à plein temps d'un mathématicien de haut niveau. D'ailleurs, quelle spécialité devrait-il choisir ?

Pire, ce mathématicien, coupé de la recherche fondamentale, verrait rapidement ses connaissances devenir obsolètes. Pour ne pas montrer à son employeur les limites de ses compétences et l'usure de celles-ci, il aura naturellement tendance à couper son entreprise des relations universitaires qui pourraient lui porter ombrage, sauf s'il est à un poste de responsabilité tel que son passé de mathématicien n'est pour lui qu'une -excellente- formation générale.

Dans ce domaine, la solution optimale pour l'industriel me paraît donc le recours au conseil extérieur qui lui permet, chaque fois que le besoin s'en fait sentir, d'avoir recours au meilleur spécialiste du moment dans la discipline concernée.

La formation mathématique pour le scientifique devant travailler dans l'industrie doit donc lui permettre :

- D'avoir conscience de l'efficacité d'un outil mathématique élaboré.

- D'avoir conscience de l'écart gigantesque entre les mathématiques de bases apprises dans une école d'ingénieurs ou à l'université, et l'outil dont disposent des chercheurs en mathématiques.

- De discerner les problèmes où les mathématiques peuvent être utiles.

- De savoir poser un problème aux mathématiciens.

Cet apprentissage ne peut se faire qu'au contact réel de mathématiques de haut niveau, quel que soit d'ailleurs le domaine.

Personnellement, je pense que la physique théorique est une excellente école en la matière. Utilisatrice de mathématiques de haut niveau dans plusieurs domaines, au carrefour de plusieurs disciplines scientifiques, elle apporte une ouverture d'esprit sur l'approche des problèmes physiques, sur la modélisation, et sur les outils de calcul.

★ ★ ★

Réponse de MICHELIN

A. COULOMBEAU

L'assistance d'un mathématicien, pour la résolution d'un problème rencontré dans la vie industrielle, du laboratoire de conception au bureau de gestion, en passant par l'atelier de fabrication, fait rarement appel à des mathématiques seules. Aussi, une formation complémentaire dans une ou plusieurs autres disciplines apparaît-elle souhaitable, notamment en : mécanique, gestion (économique, organisation...), physique. [...] .

Et même s'il est intéressant de noter un regain d'intérêt pour des formations de mathématiques pures (logique, algèbre...) dans des activités telles que la conception de systèmes d'information, ou l'étude de l'intelligence artificielle, il nous semble qu'une formation unique, très spécialisée, notamment en mathématiques, est difficilement utilisable dans une entreprise industrielle. Par ailleurs une formation trop spécifique ou trop exclusive serait dommageable à la carrière des

individus qui, sauf exception, ne souhaitent vraisemblablement pas
passer toute leur vie dans un même type d'activité trop délimitée. Dans
ce domaine comme dans d'autres, l'ouverture d'esprit la plus vaste
possible paraît souhaitable, ce qui doit trouver une certaine forme de
traduction au niveau de la formation. Toutefois, nous n'entendons pas
par là, pour ce qui concerne la formation en mathématiques elles-mêmes,
un enseignement exclusivement utilitaire : au contraire, la formation de
mathématiciens ouverts aux nouvelles voies de recherche en mathématiques
et maîtrisant les nouveaux concepts qui y sont développés procède de
cette ouverture d'esprit dont ont besoin les entreprises : c'est une des
voies du progrès. Constatons encore que les formations de mathématiques
appliquées (analyse numérique, statistique, traitement du signal)
sont encore très prisées.

TEACHING MATHEMATICS TO MATHEMATICIANS
AND NON-MATHEMATICIANS

G.A. Jones
University of Southampton, Southampton, United Kingdom

ABSTRACT

In this paper it is argued that the rigid categorisation of
mathematics teaching into the teaching of mathematics as a
subject in its own right and the teaching of mathematics as
a service subject is too clear-cut and specific. Rather
mathematics teaching should be viewed as a spectrum of
activities which are underpinned by a broad but adaptable
philosophy. Three examples are then given of subject areas
amenable to a treatment in which theory and application are
presented in a unified and mutually supportive way.

"At the age of 18 there were nice occasions in
Zürich where I was to listen to talks by higher
mathematicians like Kollross and Weyl and people
like that. I didn't get too much from them, but
anyway they were stimulating." (Marcel Golay)

1. INTRODUCTION

The ICMI discussion document [1] implicitly
distinguishes two modes of teaching mathematics in higher
education :-
(A) as a subject in its own right, and
(B) as a service subject.
Indeed, it encourages the reader to separate (B) from (A),
and to consider it in isolation. I shall argue that, while
this distinction may be valid as an aid to general
discussion, when it comes to specific cases matters are far
less clear-cut. Rather than two separate philosophies for
teaching mathematics, I would prefer to see one broad
philosophy which can be adapted continuously to meet the
needs of the students involved.

For example, many universities now offer a
proliferation of degree courses, ranging from Mathematics,
through Mathematics with X, Mathematics and X, X and
Mathematics, X with Mathematics, to X itself. Where does
one draw the boundary between modes (A) and (B)? Even
within a given degree course, those involved may have widely
differing attitudes to the status of individual topics.
Consider, for instance, a linear algebra course for first-
year mathematics students. A pure mathematician may regard
this as part of tradition (A), leading on to more advanced
concepts (functional analysis, representation theory,
etcetera), whilst a statistician or numerical analyst may
follow (B) and regard it as essentially a service course,
providing techniques for handling matrices in his own
discipline. (It is not simply a case of pure mathematicians
versus the rest here - a theoretical physicist, needing to
progress to deeper concepts from pure mathematics, may
consider (A) more important than (B) in this instance.)

The distinction becomes even more blurred when one
considers the wide variety of careers open to graduates:
thus an economics student may become a professor of
statistics in a mathematics department, while a specialist
in pure mathematics may go on to do research in electronics
on program verification.

The difficulty for the teacher, then, is that, whereas
his course may be formally designated as of type (A) or (B),
his audience may bring to it a variety of backgrounds,
abilities, interests and expectations. The danger is that,
by automatically basing his style of teaching on some
version of philosophy (A) or (B), he may force his students
into the stereotypes of producers or consumers of
mathematics, two groups which are finding it increasingly
difficult to communicate with each other.

In reality, there is a continuous spectrum of
approaches to and applications of mathematics, and many of
the most successful practitioners are those who are able to
move around, finding useful ideas and techniques in
unlikely-looking places. For example, consider how Cormack
used the Radon transform to solve problems in tomography, or
how Donaldson borrowed instantons from theoretical physics
to create new structures in differential topology (see [2],
pp. 716-733 and 900-903 for informal accounts of these and
other achievements). Of course, we cannot expect all our
students to win Nobel prizes or Fields medals, but at least
we can try to give them a feeling for the breadth of
mathematics, and the confidence to explore it, rather than
simply training them to work in one particular area.

2. SOME EXAMPLES

A possible method for achieving these objectives is to
combine theory and practice in teaching, and to emphasize
links between areas of mathematics and its applications. A
classic example is Courant's book 'Differential and Integral
Calculus' [3], which combines an account of the techniques
of calculus with superbly written digressions into its
analytic foundations and physical applications.

At higher levels, it is often possible to focus more directly on specific areas where theory and practice have converged. One can use books and papers, not just as sources of additional information, but also to illustrate how authors have (with varying degrees of success) tried to communicate mathematically with different types of readership. This approach can be particularly valuable if the course is attended by students from different disciplines: ideally, each student will find at least one acceptable account of the main topics, and if students are encouraged to work together then their different insights will reinforce each other.

To illustrate this, I will discuss three topics which arise in a third-year course I teach on Information and Coding Theory, aimed at students specialising in Mathematics, Computer Studies or Electronics. For our purposes, this is an ideal subject, having been founded (less than 40 years ago) by engineers and by mathematicians working in an engineering environment. The early papers tend to be short and not too technical (and sometimes not too rigorous); later, a more "professional" approach takes over, and the papers become more mathematical and (with a few honourable exceptions) less readable. Most of the textbooks are either too mathematically sophisticated for the non-mathematicians on the course, or are too informal for the mathematicians; those which seem to find a good balance, such as [4,5], tend to concentrate on, say, coding theory at the expense of information theory. Hence the preparation of this course involved both supplying rigour for intuitive arguments, and simplifying proofs which were correct but too technical. It was not an easy task, but I learnt a great deal more about the subject than if I had simply chosen a textbook and followed that. I hope that the students have also gained from this approach.

The three topics I have chosen are as follows :-

3. THE SARDINAS-PATTERSON THEOREM

First we need some definitions. Take an alphabet (finite set of symbols) A, and let A^+ be the set of all (non-empty) <u>words</u> in A; these are finite sequences $w = a_1 \ldots a_r$ of elements of A. A <u>code</u> is a non-empty finite subset $C \in A^+$, and a <u>string</u> $s = w_1 \ldots w_k$ is formed by juxtaposing finitely many codewords $w_i \in C$ (without punctuation or separation by blanks). One says that C is <u>uniquely decodable</u> (or simply u.d.) if each such <u>s</u> arises from a <u>unique</u> sequence of codewords, that is, if

(3.1) $w_1 \ldots w_k = v_1 \ldots v_m$ $(w_i, v_j \in C)$
                              $\Rightarrow k = m$ and each $w_i = v_i$

     For example, if $A = \mathbf{Z}_2$ then the code $C = \{0, 10, 11\}$ is u.d. (simply read any string <u>s</u> from left to right), whereas $C' = \{10, 101, 110\}$ is not (since $\underline{s} = 10110$ arises in two ways). In many cases, such as $C'' = \{0010, 011, 01101, 1001\}$, the answer is not obvious.

     In 1953, Sardinas and Patterson [6] provided a necessary and sufficient condition for a code to be u.d. Changing their notation a little let $C_0 = C$, and for integers $n \geq 1$, let $C_n = \{w \in A^+ | uw = v$ where $u \in C, v \in C_{n-1}$, or $u \in C_{n-1}, v \in C\}$. Let $C_\infty = \bigcup_{n \geq 1} C_n$. Then C is uniquely decodable if and only if:

(3.2)   C and C_∞ are disjoint sets.

     The Sardinas-Patterson paper is short and readable. First there is an informal description of the problem, with examples from natural and artificial languages. Then there is a statement of criterion (3.2), with a few illustrative examples. The next part of the paper is more mathematical in format: definitions and theorems are stated in the language of predicate calculus, but each is immediately followed by an informal interpretation; proofs are omitted, the reader being referred to an unpublished research report for details. Finally, the authors pose some open problems: for example, is their test an <u>algorithm</u>, in the sense that it will terminate? If C_∞ contains a codeword, one will discover this in a finite amount of time by computing C_1, C_2, ... in succession, but if C_∞ contains <u>no</u> codewords, does one have to compute <u>every</u> C_n to discover this? (In fact, it is not difficult - and much easier than proving the main theorem - to show that the test is indeed an algorithm: by induction on n, each $w \in C_n$ $(n \geq 1)$ is shorter than the longest codeword, so only finitely many subsets of A^+ can arise as C_n for any n; thus the sequence of sets C_1, C_2, ... is eventually periodic, so the computation of C_∞ is a <u>finite</u> task.

Subsequently, several books and papers have included proofs of the Sardinas-Patterson Theorem, at various levels of rigour and completeness. Bandyopadhyay [7] proves that the condition (3.2) is necessary, but merely gives an example to illustrate its sufficiency, whereas Seeley [8] proves this; both papers are fairly straightforward. Riley [9] uses rather more elaborate formalism to give a proof which is "a good deal simpler" (once one has mastered his notation); he also proves related results of Even and Levenshtein on delay and synchronizability of codes. Ash's book [10] gives a detailed proof of the theorem, Gallager [11] presents the formulation of an equivalent version of the test as an exercise, while Even [12] justifies Gallager's test and considers it from an algorithmic point of view.

One can also consider unique decodability in terms of abstract algebra. The set A^* of <u>all</u> words in A (that is, A^+ together with the empty word) is a monoid under juxtaposition: it has an associative binary operation, with a 2-sided identity element. Algebraists will recognise A^* as the free monoid with basis A, and (3.1) as the condition that the submonoid of A^* generated by C is free, with basis C. This insight, due to Schützenberger [13], has lead to an algebraic treatment of codes, similar in spirit to certain branches of combinatorial group theory, but also having links with the theories of automata and formal languages; see, for instance, the books by Berstel and Perrin [14] and Lothaire [15].

I have considered this example in some detail in order to show how a single topic can be approached from many points of view, and can be described in a wide variety of contexts and degrees of rigour. In a specific situation, a teacher must choose which treatment of a topic is most appropriate for his audience, but he is doing them a disservice if he totally ignores alternative treatments and contexts: for many students, the main difficulty in studying mathematics is not so much that of understanding individual topics, as that of seeing how different topics are related to each other, and here a teacher with a wide knowledge of the literature can be of great help.

For the lecture-course I have described, my preferred method of teaching this topic is:

1) a statement of the theorem, with a reference to the original paper, and a few examples like C, C′ and C″ above to provide motivation and to clarify the concepts involved;

2) an outline proof that (3.2) is necessary, using C″ to show how, if some C_n contains a codeword, one can construct a situation where (3.1) is violated;

3) brief comments on the converse, and on how this theorem is related to other topics not covered in the course;

4) some routine exercises (for the students) in applying the theorem, and some deeper exercises to provoke further insight, e.g. the proof that the theorem is an algorithm;

5) a duplicated hand-out, expanding (2) into a detailed proof and giving suggestions for further reading to support the comments in (3).

For a different audience, of course, one might prefer a different approach. Obviously, not every topic in a course can be treated in this great detail; however, many courses can be seen as depending on a handful of major results, and at least some of these could be presented in this way. Two more examples follow.

4. HUFFMAN CODES

If the codewords w_i of a code C have wordlengths l_i, and are used with relative frequencies p_i ($p_i \geq 0$, $\Sigma p_i = 1$), then the <u>average wordlength</u> of C is L(C)= $\Sigma p_i l_i$. Given an alphabet A and a probability distribution (p_i), an <u>optimal</u> code is a u.d. code $C \subseteq A^+$ which mimimises L(C). In 1952, Huffman [16] gave an algorithm for constructing such codes, by systematically associating longer codewords with lower probabilities p_i (Morse code does this rather less systematically). This algorithm can be thought of as producing a rooted tree with minimum weighted path length;

as such, it also has applications in computer programming, e.g. optimal search procedures ([17], §2.3.4.5, and [18]) and sort-by-merge techniques ([12], §4.3).

Huffman's paper is easy for students to read: it has a clear, informal description of the algorithm, and some good examples. However, his proof of the optimality of the resulting codes (now called <u>Huffman codes</u>) is somewhat sketchy: he derives some necessary conditions for a code to be optimal, and then simply asserts that they are also sufficient, a fact which is not entirely obvious. Moreover, although he notes the minor modification his algorithm requires in the non-binary case (i.e. when $|A| > 2$), he gives no hint that the proof is rather more complicated here.

Some textbooks [10,11,12,17,19] give fuller proofs, but none is totally satisfactory. For example, the standard proof that a Huffman code C is optimal is to choose an optimal code D, and then use Huffman's necessary conditions on D to show that $L(C) \leq L(D)$; however, this implicitly assumes that optimal codes exist, a fact which requires proof since there are infinitely many u.d. codes and their average wordlengths might not <u>attain</u> a greatest lower bound. This difficulty can easily be resolved (far more easily than the corresponding Dirichlet principle in potential theory), but I know of no book or paper which even mentions this point. In teaching, one may not always wish to confront such technical details directly, but one should at least indicate that problems exist.

5. PERFECT CODES

In the theory of error-correcting codes, <u>block</u> codes are used, those in which all words have the same length n, so that $C \subseteq A^n$. To minimise the effect of errors in transmission, one looks for codes C with a large <u>minimum distance</u>

$$d = \min \{ \, \delta(u,v) \mid u,v \in C, \ u \neq v \}$$

where $\delta(u,v)$ is the <u>Hamming distance</u>, the number of coordinate places in which u and v differ. To transmit information at a suitably high rate, one also wants $M = |C|$

to be large. A simple counting argument due to Hamming [20] shows that

(5.1) $M(1 + \binom{n}{1}(q-1) + \ldots + \binom{n}{t}(q-1)^t) \leq q^n$,

where $t = \lfloor \frac{1}{2}(d-1) \rfloor$ (the number of errors corrected by C), and $q = |A|$. <u>Perfect</u> codes are those which attain equality in (5.1); they include the two <u>Golay codes</u>, for which $(n,q,M,d) = (11,3,3^6,5)$ and $(23,2,2^{12},7)$ respectively. These, and other examples of perfect codes, were described in a remarkable half-page paper by Golay [21] in 1949. Although Golay had little formal training in mathematics, his interest in the subject (e.g. in number theory) was captured at an early age, as shown by the quotation I have used at the beginning of this paper. This is taken from Thompson's fascinating book [22] on the early work of Hamming and Golay on perfect codes; Thompson also describes how Leech subsequently used the Golay code of length n = 23 to construct a lattice Λ corresponding to an exceptionally dense packing of spheres in $\mathbb{R}^{24}$, and how Conway then used the symmetry group of Λ to construct the three sporadic simple groups which bear his name. Other structures intimately related to the Golay codes include Steiner systems, Hadamard matrices, and the Mathieu groups [4,5,23].

Although Golay's paper is very precisely-written, it is extremely terse, and gives little hint of how he discovered his codes; nevertheless, taken together with Hamming's more leisurely paper and Thompson's very thorough historical investigation, it can be used to illustrate a remarkable sequence of mathematical discoveries. For further background on perfect codes, including the work of van Lint and Tietäväinen towards their classification, see [24,25].

6. CONCLUSION

I have used the three examples above to illustrate how one can teach a course to mathematicians and non-mathematicians, by focussing on a few major topics, and by using the literature to point students towards other branches of mathematics and its applications. One can also introduce original papers for their historical significance, and for their ability to draw together the different strands

that make up a course; thus no conscientious teacher would
do as I have just done, and discuss information and coding
theory without referring to the paper of Shannon ([26],
reprinted in [27]) which originated the subject. The
essential point is to present a sufficient variety of
approaches that each student can find something appropriate
to his interests and abilities.

REFERENCES

1. A.G.Howson et al, Mathematics as a service subject,
    L'Enseignement Mathématique, 32, 1986, pp 159-172.
2. Notices American Mathematical Society, 33, 1986.
3. R.Courant, Differential and integral calculus, Blackie,
    1934.
4. R.Hill, A first course in coding theory, Oxford
    University Press, 1986.
5. V.Pless, Introduction to the theory of error-correcting
    codes, Wiley, 1982.
6. A.A.Sardinas and G.W.Patterson, A necessary and
    sufficient condition for unique decomposition of coded
    messages, Institute of Radio Engineers International
    Convention Record, 8, 1953, pp 104-108.
7. G.Bandyopadhyay, A simple proof of the decipherability
    criterion of Sardinas and Patterson, Information and
    Control, 6, 1963, pp 331-336.
8. D.A.R.Seeley, A short note on Bandyopadhyay's proof of
    the decipherability criterion of Sardinas and Patterson,
    Information and Control, 10, 1967, pp 104-106.
9. J.A.Riley, The Sardinas/Patterson and Levenshtein
    theorems, Information and Control, 10, 1967, pp 120 -
    136.
10. R.Ash, Information theory, Interscience, 1965.
11. R.G.Gallager, Information theory and reliable
    communication, Wiley, 1968.
12. S.Even, Graph algorithms, Pitman, 1979.
13. M.P.Schutzenberger, Une théorie algébrique du codage,
    Séminaire Dubreil-Pisot, exposé 15, 1955.
14. J.Berstel and D.Perrin, Theory of codes, Academic Press,
    1985.
15. M.Lothaire, Combinatorics on words, Addison-Wesley,
    1983.
16. D.A.Huffman, A method for the construction of minimum -
    redundancy codes, Proceedings of the Institute of Radio
    Engineers, 40, 1952, pp 1098-1101.
17. D.E.Knuth, The art of computer programming I:
    Fundamental algorithms, Addison-Wesley, 1968.

18. S.Zimmerman, An optimal search procedure, <u>American Mathematical Monthly</u>, 66, 1959, pp 690-693.
19. R.J.McEliece, <u>The theory of information and coding</u>, Addison-Wesley, 1977.
20. R.W.Hamming, Error detecting and error correcting codes, <u>Bell System Technical Journal</u>, 29, 1950, pp 147-160.
21. M.J.E.Golay, Notes on digital coding, <u>Proceedings of the Institute of Radio Engineers</u>, 37, 1949, p 657.
22. T.M.Thompson, <u>From error-correcting codes through sphere packings to simple groups</u>, Mathematical Association of America, 1983.
23. I.Anderson, <u>A first course in combinatorial mathematics</u>, Oxford University Press, 1974.
24. J.H.van Lint, <u>Introduction to coding theory</u>, Springer, 1982.
25. F.J.Macwilliams and N.J.A.Sloane, <u>The theory of error-correcting codes</u>, North-Holland, 1976.
26. C.E.Shannon, A mathematical theory of communication, <u>Bell System Technical Journal</u>, 27, 1948 pp 379-423 and pp 623-656.
27. C.E.Shannon and W.Weaver, <u>The mathematical theory of communication</u>, University of Illinois Press, 1963.

THE MATHEMATICS SERVICE COURSES ENVIRONMENT

E.R. Muller
Brock University, Ontario, Canada

B.R. Hodgson
Université Laval, Québec, Canada

ABSTRACT

Mathematics service course work is widespread and diverse in the majority of Canadian universities. This service is provided in a context of open admission which results in a heterogeneous mathematics preparation. Differences in mathematics preparation are accentuated by the importance of adult and continuing education where a large proportion of the students have been out of school for a number of years. This paper studies the serious response developed by some mathematics faculty and departments. It is recommended that more effort be placed on the use of appropriate software in this important rôle played by mathematics.

1. INTRODUCTION

This paper assumes the general service course situation as it exists in Canada [1] and discusses the specific situation as it is experienced in departments of mathematics in many Canadian universities. This situation can be characterized by

(i)   a variety of service courses offered for students in administration (accounting, marketing, etc.), agriculture, biology, chemistry, computer science, economics, engineering, geology, medicine, physics, etc., etc.,

(ii)   a mathematics department composed of a majority of faculty members who would prefer not to be involved with service course work,

(iii)   a chairperson who must allocate teaching loads carefully balancing the external and internal pressures,

(iv)   an open admission producing a heterogeneous secondary school mathematics preparation, accentuated by a significant number of students who have been out of school for differing amount of time.

It is within this context that we shall address, in Section 2, the responsibilities of and subsequent implications to the mathematics department; this we follow by some discussion in Section 3 on the effective implementation of service courses. Section 4 considers motivation and innovations in service course work. In Section 5, we comment on the use of computers in service courses which, we believe, will be of greater and greater importance as the departments which are serviced make more use of these systems.

2. SERVICE COURSES: RESPONSIBILITIES AND IMPLICATIONS

Students in mathematics service courses often perceive mathematics as one of the following:

(i)   a low priority in their overall academic program,

(ii)   of little initial relevance to their major subject,

(iii)   a hurdle to be cleared -- perhaps placed to limit entrance to their chosen field or professional school,

(iv)   a subject they had hoped to avoid at university.

These students lack motivation in mathematics and the courses are seen as trials to be endured, usually with a large number of other students. Their aim is to pass the courses with a minimum amount of work; however they have concern that the grade might lower their overall average.

The mathematics department usually consists of a majority of faculty members who would prefer not to be involved with service course work and a chairman who must allocate teaching loads carefully balancing the external and internal pressures. Many of the mathematics faculty members perceive a service course as one or more of the following:

  (i)   a majority of uninterested students,
  (ii)   a large class -- more student contact with faces one recognizes but names one forgets,
  (iii)   a class containing "general" students with rather weak mathematics preparation, faced with a syllabus too difficult for them, thus likely to reinforce their previous disastrous mathematics school experiences,
  (iv)   a restricted and overloaded syllabus,
  (v)   a techniques oriented emphasis,
  (vi)   a list of topics remote from traditional research interests in mathematics,
  (vii)   the possibility of evaluation from outside the department,
  (viii) a high administrative load,
  (ix)   a task where little university credit is given for generally less interesting yet more time consuming duties.

Even when a mathematics department has a strong desire to give service courses of a high quality, it is not always easy to see how this can be achieved. For instance, some host departments might have difficulty in identifying their mathematical needs; other are not willing to make the effort to assess these needs, preferring to use mathematics as a filter for their students. At the other extreme, one finds professors in other disciplines who have a very clear view of the kind of mathematics they want their students to take, but who would much rather teach it themselves than leave it in the hands of mathematicians (some might argue that such a position stems from the desire of these teachers to better motivate their students -- in the spirit of the commentaries in Section 2.5.1 of [2] -- while others could interpret it as aiming to control what is being taught to their students, independently of external input from a mathematics department). In all these situations, mathematics departments involved in service teaching are in the delicate position of negotiating contents of courses with various programs which have very different mathematical needs.

This might explain the great difficulties often encountered in trying to construct a coherent sequence of service courses in mathematics with clearly identified objectives. In addition to the problem of coping with students entering university with very different mathematical backgrounds (placement tests are often useful in this respect for guiding students into some remedial courses), there are all the various constraints imposed by the host departments which prove to be far from easy to satisfy. A typical such constraint might be described as "squeezing" a sometimes unreasonable amount of mathematical content into a severely restricted number of academic credits. Some professional programs will tend to justify their position by referring to policies of professional associations. For example, the Canadian Council of Professional Engineers has a standing committee, the Canadian Accreditation Board, whose task is to unify the methods and procedures of accreditation of the educational qualifications of graduates from all engineering programs across Canada. This Accreditation Board has stipulated (see [3], p. 19) that an accreditated program, based on a normal four-year Bachelor's degree program, should include a minimum equivalent to one half year of "mathematical foundations" (matrix algebra, differential and integral calculus, differential equations, probability and statistics, numerical analysis). The fact is that for most engineering programs, this minimum has also become a maximum so that all mathematical knowledge for these students has to be accomodated within the equivalent of one half year. This creates a lot of pressure not only on the allowable number of mathematical courses and their content, but even on the number of academic credits given to a particular course (if a mathematics course is given, say, 2 credits instead of 3, this leaves one extra credit to be assigned to some other engineering course). In other cases host departments will be interested in some advanced topics for their students, for instance numerical resolution of systems of linear equations, but will be unwilling to require the necessary prerequisites in linear algebra.

The emergence of numerical and symbolic software places an additional pressure on mathematics departments. Many know-hows conveyed in mathematical service courses are now readily available on computers. Sophisticated numerical analysis packages have been available for more than a decade but the impact on undergraduate mathematics teaching of symbolic manipulation packages such as MACSYMA, Maple or even muMATH is still to be felt. A report of teaching engineering analysis using such systems at Cornell University is given in [4]. Some host departments are now asking whether the basic rôle actually played by mathematics in the formation of their students could be better replaced by some informatics training. Could a substantial part of the work actually done by mathematicians in the preparation of scientists of all disciplines be replaced by more "relevant" courses offered by computer science departments? We do not support this view and believe that mathematics departments must put in place service courses that stress the long run superiority of a strong theoretical background for the understanding of models, whether or not they are implemented on the computer (see also [5]). Mathematicians must demonstrate that their teaching and their courses are aimed at the formation of scientists, and not mere technicians. On the other hand if mathematics departments do not educate students in the proper use of such software, they can expect to loose service course enrolments to computer science departments which are experiencing drop in enrolments (see [6] for recent data on computer science enrolment in Canadian universities).

3. EFFECTIVE IMPLEMENTATION OF SERVICE COURSES

How does a mathematics department successfully meet the diverse objectives of the host department, the sometimes less than enthusiastic faculty and the large enrolments? We believe that this is facilitated by good course management which involves

1. Consistency between various sections within the one year and also from year to year. This is not always easy to achieve. It requires instructors to meet well ahead of the scheduled offering to
   (a) select text(s) which reflects the emphasis of the host department(s),
   (b) set a week by week schedule of lecture content,
   (c) agree on evaluation procedures for all sections of a given year which are comparable to those used in previous years.

2. Providing the student very early in the course, preferably at the first meeting,
   (a) the detailed course outline which crossreferences the text according to weekly lecture content,
   (b) the complete set of problems for the course with detailed references to course materials, a list of due dates for materials to be handed in, or suggested due dates if there are no assignements,
   (c) the complete evaluation procedure with weights and penalties.

Some faculty have argued that this course management spoon feeds the students. We would counter that we are providing the students all the details of a course with a rigid syllabus where enrolments and circumstances provide little if no flexibility. Within this rather rigid structure there is still room for motivation and innovation, as we explore in the next section.

4. MOTIVATION AND INNOVATIONS IN SERVICE COURSES

We can report on a number of constructive ideas which have been implemented in service courses. In all cases we list the advantages and disadvantages which were experienced on implementation.

1. Topics were introduced in a statistics course to coincide as closely as possible with their need in a laboratory course in biological sciences (appropriate biological labs were crossreferenced in the statistics notes).
   Development of algorithms in an introductory discrete mathematics course were coordinated with the development of the programming language in a computer science course (see [7]).

Advantages
(a) Reinforcement of concepts.

(b) Relevance and motivation.

Difficulties
(a) Requires close cooperation between departments.
(b) Requires advance planning and course changes must be coordinated.

2. Selecting a required text oriented towards the student's major discipline, and selecting mathematics problems oriented towards that discipline.

Advantages
(a) Demonstrates the "raison d'être" for the mathematics course.

Difficulties
(a) Texts are often not available.
(b) Students must overcome not only the difficulty of mathematics but also the difficulty of understanding the situation where mathematics is to be applied.
(c) Many of the "applied" problems are artificial.
(d) The mathematics faculty must have a good understanding of the other discipline.
(e) Many service courses are not composed of majors in a single discipline.

3. Generating original and individualized data for each student. Although this is most easily done for statistics and operations research courses, some variation in questions and constants have been used in large enrolment courses in calculus (see reference [8]).

Advantages
(a) Students are encouraged to work together and help each other.
(b) For evaluation purposes one can put more emphasis on term work and less on tests and examinations.

Difficulties
(a) Very time consuming to set up and maintain computer programs.

4. Use of Case Studies. This method of teaching is gaining popularity in the social sciences. It has proved very rewarding both for faculty and students in upper level service courses.

Advantages
(a) Immerses the student in applications of mathematics to his/her discipline.
(b) Students work together as a team.
(c) When results of the case studies are presented other students can be involved in the evaluation.

Difficulties
(a) Not suitable for classes with large enrolments.
(b) It is difficult to find good cases.
(c) Normally not a situation where students are exposed to mathematics they did not know (which is counter to what many service courses are presently structured to do -- namely expose the student to a long list of mathematical topics).

There are a number of operations research texts which are devoted to Case Studies. The UMAP Journal covers a broader set of mathematical applications as does the research journal Mathematical Modelling (it should be pointed out that a special issue of the latter is devoted every year to the best papers presented to the Mathematical Competition in Modelling, an annual contest for undergraduate student teams organized since 1985 by the Consortium for Mathematics and its Applications (COMAP)). The National Council of Teachers of Mathematics has recently brought out two books [9, 10] on applications at a lower level.

5. USE OF COMPUTERS IN SERVICE COURSES

Earlier it was mentionned that the emergence of numerical and symbolic computer software is placing an additional pressure on mathematics departments. Results of our survey reported in [1] demonstrate that some mathematics departments have responded to the general availability of statistical software. We have no such evidence that symbolic mathematical software is being used in service courses in areas of calculus, linear algebra, etc.

How is symbolic mathematical software to be incorporated into service courses? From a pedagogical point of view these systems have much to offer (see reference [11]):
   (a)   a rich environment for exploration at different levels of generalization,
   (b)   motivation and enrichment,
   (c)   the ability to generate a large number of examples and exercices,
   (d)   the exposure to more realistic problems to develop a sense of reasonableness, estimation, etc.
On the other hand there is a great concern about introducing into first year calculus classes a micro (or even a hand-held calculator!) which can do the majority of textbook problems presently used in these courses.

We believe that it is vital that more mathematics departments begin experimenting with the introduction of symbolic mathematical software in their service courses and that the results of these experiments be communicated to others. Already many users of mathematics have shown some eagerness that their students learn to use these systems in working on the mathematical models for their problems in their own fields (see for example in [12] some papers on symbolic computation in the education of chemistry students; these papers were presented to a symposium organized by the American Chemical Society's Division of Computers in Chemistry). The report of the Tulane Conference "To Develop Alternative Curriculum and Teaching Methods for Calculus at the College Level" [13] presents a number of different suggestions for changes in calculus courses, some of these applicable in the service course area.

It appears to us that ICMI is the appropriate body to motivate research and disseminate the results of the introduction of symbolic mathematical systems in mathematics service courses.

REFERENCES

[1] Hodgson, B.R. and E.R. Muller, "Mathematics service courses: a Canadian perspective." In: Clements, R.R. and A.G. Howson (eds.) *Selected Papers on the Teaching of Mathematics as a Service Subject.* (ICMI Symposium, Udine, 1987) Springer-Verlag.

[2] Howson, A.G., J.-P. Kahane, P.J. Kelly, P. Lauginie, T. Nemetz, F.H. Simons, C.A. Taylor and E. de Turckheim, "Mathematics as a service subject." *Enseign. Math.* 32 (1986) 159-172.

[3] Canadian Council of Professional Engineers, *1985 Report of the Canadian Accreditation Board.*

[4] Lance, R.H., R.H. Rand and F.C. Moon, "Teaching engineering analysis using symbolic algebra and calculus." *Engin. Educ.* 76(2) (1985) 97-101.

[5] Hodgson, B.R., "Symbolic and numerical computation: the computer as a tool in mathematics." In: Lovis, F.B. and D.C. Johnson (eds.) *Informatics and the Teaching of Mathematics.* (Proc. IFIP TC3/WG3.1 Conference, Sofia, 1987) North-Holland, to be published.

[6] White, L.J., "Computer science enrollment at Canadian universities." *Abacus* 4(3) (1987) 37-40.

[7] Jenkyns, T.A. and E.R. Muller, "Discrete mathematics: two years experience with an introductory course." In: *The Influence of Computers and Informatics on Mathematics and its Teaching.* (Supporting Papers for the ICMI Symposium, Strasbourg, 1985) IREM, Université Louis-Pasteur, Strasbourg, 1985, pp. 247-250.

[8] Auer, J.W., T.A. Jenkyns, C.F. Laywine, J.P. Mayberry and E.R. Muller, "Motivating non-mathematics majors through discipline-oriented problems and individualized data for each student." *Int. J. Math. Educ. Sci. Technol.* 13 (1982) 221-225.

[9] Sharron, S. and R.E. Reys, *Applications in School Mathematics.* National Council of Teachers of Mathematics, 1979. (1979 Yearbook)

[10] Bushaw, D., M. Bell, H.O. Pollak, M. Thompson and Z. Usiskin, *A Sourcebook of Applications of School Mathematics.* National Council of Teachers of Mathematics, 1980.

[11] Hodgson, B.R. and E.R. Muller, "Symbolic mathematical systems and their effects on the curriculum." *Notes Canadian Math. Society* 18(6) (1986) 26-33.

[12] Pavelle, R. (ed.) *Applications of Computer Algebra.* Kluwer Academic Publishers, 1985.

[13] Douglas, R.G. (ed.) *Toward a Lean and Lively Calculus.* (Report of the Conference/Workshop To Develop Curriculum and Teaching Methods for Calculus at the College Level, Tulane University, 1986) Mathematical Association of America, 1986.

MATHEMATICS AS A SERVICE SUBJECT
AN INTERACTIVE APPROACH

N. Patetta
Universidad CAECE, Buenos Aires, Argentina

ABSTRACT

The present paper is a contribution for the call for papers required by
the ICMI about the ICMI's study "Mathematics as a service subject".
It takes into account the results of an inquiry performed in the Mar del
Plata University by M.I.Aguirre, M.G.Monterrubianessi and G.Pellicer,
members of the Seminar about the teaching of Mathematics that took place
in Mar del Plata the first semester of 1986.
The aim of the paper is to make evident the rejection towards the
mathematical teachers observed in students of other disciplines and to
propose a solution by means of an interactive action between mathematical
teachers and the specific teachers of each discipline.

Mathematics as a service subject is a hard theme today. A permanently increasing number of disciplines demand mathematics to be used in their subjects and each of them claims for persuading and illuminating ways of teaching.

If we ask:

Why do we teach mathematics to the students of discipline X?

Probably we could find, in a first approach, a naive answer:

Because discipline X needs mathematics
to explain its proper subject matter.

But, not necessarily the mathematical necessities of discipline X coincide with the feeling that the students have of these necessities.

In fact, we can observe different feelings among the students according to the level of knowledge about discipline X that has been got by them.

For instance, in an extreme point, we have students in a post graduate level.

The mathematics courses taught to post graduate students in any discipline are usually pleasant for the mathematics teachers.

They teach to interested people that actually need the courses to understand and to solve problems in their disciplines and the mathematical knowledge taught is absolutely necessary for the students advances in their disciplines, and what is most important, the students are aware of it.

Because of this real necessity the students are interested in any theme, with all their capacities of learning put in the courses. Any special motivation by the mathematics teachers is unnecessary. The students are sufficiently self motivated and they suggest to the mathematics teachers the applications of the different topics.

They enjoy recognizing the mathematics that is just needed for their present problems. Usually pure mathematicians are required to teach in this kind of courses.

But, unluckily, the most of the teaching work is fulfilled in the undergraduate courses, and in this level the situation is very different.

First of all, we must take into account that the mathematical courses together with the courses of basic sciences are usually taught in the first years of study, previous to the specific courses of each discipline.

For this reason, the students do not know enough about their proper discipline at the moment of taking the mathematics courses; therefore they can not use illuminating examples of application.

We can add that if the teachers of the specific subjects do not help enthusiastically to incorporate mathematics (using it in their courses!) a natural rejection is generated in the students. They will have a lack of confidence in their mathematics teachers, and they will prefer to learn mathematics with teachers of the major discipline and not with mathematics teachers.

They have the feeling of wasting their time learning with mathematics teachers (They suppose that only the specialists in discipline X know the "useful things").

It is very interesting to observe the difference of attitudes in the rejection towards the mathematics teachers according to the intensity and the way that mathematics is used in the different disciplines.

In an inquiry performed among students and graduates of different disciplines in the Mar del Plata University (Argentina) one of the questions was:

>Who, do you think, must teach the
>mathematics subjects in your courses?

The three possible answers were:

>a- teachers of mathematics
>b- teachers of discipline X with
>   mathematical knowledge
>c- a team of both.

The results according to the discipline were the following:

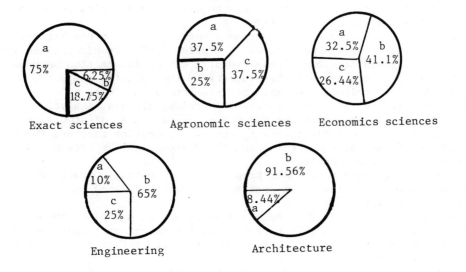

Exact sciences   Agronomic sciences   Economics sciences

Engineering   Architecture

It is possible to observe an increase in the rejection towards the Technical and Arts studies.

Perhaps the previous description ("principle of rejection") is valid in a wide range of universities and countries in a general sense, but its intensity is associated with the regional context, more precisely, with the degree of development that the country has.

As an example, we can consider the Engineering in a non developed country. In this context, it is very strange to find engineers who can work in engineering design (only perhaps in some academical centres).

Practically all the engineers produced by the universities must work either in administrative roles or in a technical work that is reduced to having a good knowledge and handling of technical handbooks.

The students are aware of this situation and this explains their claim for studing only "the useful things".i.e. only the questions that are effectively used in the standard work. They have an important argument:

Why must we study all these abstract questions that nobody uses? (and which are usually difficult to learn).

Other complications are introduced by the educational policy of the country, specially if it is an open policy. This is the case when the universitary studies are considered a right for all the people with a non-restricted access to the University, independently of the necessities of the country (That is the Argentina case).

This policy produces an overpopulation in the classes with students without a definite vocation for their studies.

All this description leads us to the affirmation that the problem of which mathematics to teach, and how to do it, is not possible to be solved in the only context of mathematical courses. It is necessary to solve it taking into account all the subjects of the course of studies.

It is necessary not only to teach the mathematics topics but also to motivate the students to learn them.

A brief description of an experience began in 1986 in the Mar del Plata University (when a new course of studies in Electronic Engineering was opened), could be a useful example of a way of interaction between the mathematics teachers and the teachers of the specific subjects.

In the new curricula the basic and elementary knowledge of mathematics (differential and integral calculus, algebra, analytic geometry, etc) is taught in the first and second years of studies in four subjects. These subjects are taught by mathematics teachers (some of the auxiliaries are Engineers) and the Mathematical Department is responsible for these subjects.

The advanced topics, such as Graph Theory, Stochastic Processes Transforms, etc, are taught in specific subjects organized by the Electronic Department.

For instance, Stochastic Processes is taught in a subject about theory of signals headed by a professor of the Electronic Department, but the Mathematical Department provides an associate professor to teach the mathematical aspects.

Interdisciplinary seminars between teachers of both Departments will have to be organized to assure an adequate coordination of all the team. All this organization of the curricula and activities is hoped to produce an interaction between the mathematical teachers and technical teachers. This interaction will increase the appropiate use of the mathematical aspects by the technical teachers improving the theoretical approach in the specific subjects and the mathematical teachers will acquire a reasonable knowledge of the applications of the mathematics in the specific discipline (Electronic).

In this way the students in the basic courses of mathematics will learn with mathematical teachers who will be able to motivate their classes and in the technical courses the students will actually learn the technical subjects with an adecuate mathematical background, improving the theoretical approach.

An indirect hoped result is to increase the confidence of the students in their mathematical teachers.

Obviously some years will be necessary to evaluate the results of this experience, but it is interesting to mention that an experience of interaction similar to this one has been taking place in the Faculty of Agronomic Sciences for the last ten years and it is possible to observe a suggestive percentage of acceptance of the mathematics teachers by the students in the above mentioned inquiry.

As a synthesis of all the discussion the principal idea to be remarked is:

The problem of how to teach mathematics, what mathematics must be taught and who must teach it in the discipline X is not likely to be solved only in the context of the courses of mathematics. It must be solved in the context of all the course of studies of the discipline X and to get some solution an interaction among all the teachers (mathematica and specific teachers) is necessary.

MATHEMATICS AS A SERVICE SUBJECT — THE USERS' POINT OF VIEW

D.K. Sinha
University of Calcutta, Calcutta, India

ABSTRACT

That users of mathematics, apart from those in academic establishments, can have decisive views in shaping the contours of mathematical education is the chief interest of this paper. The societal aspects of mathematics vis a vis emerging facets of mathematical sciences are yet to figure prominently in any serviceable structure of mathematics. Such issues are discussed here with a view to bringing to the fore the role of mathematics in the context of changing contemporaneity.

That mathematics is serviceable scarcely needs any
reiteration and that it has continued to be so since the
initial years of ancient wisdom is now a historical reality.
Notwithstanding its esoteric character, its enigmatic
veneer, its being abstract (using the term abstract here in
a somewhat pejorative sense), its usefulness, often even in
a limited way, has stood out as something distinctive in the
course of history. The enormous development of mathematics
has enabled its practitioners to put it into some usable
forms. The service character of mathematics has, thus,
evolved over the years. The spate of development in
contemporary history has been so pronounced that one can ill
afford to miss new dimensions of mathematics as a service
subject. The purpose of this paper is to focus on one of
such emerging aspects.
    It is trite to harp on the growing domain of clientele
for mathematics. However arid the subject may appear to be,
it has become a force to reckon with, if one is really
serious about making use of it as a powerful tool in
understanding the real world and solving practical problems.
The conflict between the abstract view of mathematics and
its ability to grapple with real and concrete problems of
life remains, doubtless, a vexed issue. Without going into
the polemics, one can safely say that the species, namely
users of mathematics, have come to stay and do wield
influences, in a substantial way, in shaping the contours of
mathematics and its instruction too. While such users
abound in academic settings, those outside have, of late,
contributed in no insignificant way to the totality of
development of mathematical enterprises.
    People in such establishments make use of mathematics
and, obviously, matter a good deal in respect of the
employment-scenario in industries, commerce, trade, etcetera
and facilitate the creation and sustenance of what may be
called mathematical manpower. The experience of the present
author, based on findings in respect of employment potential
in such enterprises, is far from being depressing, vide
Sinha [1], [2]. On the other hand, it may appear surprising
to find many enlightened employers looking for qualities in
a mathematician; for example, command of language so as to
absorb what users say and to translate it back to them,
after resolution, in language intelligible to users. One of
the limitations of such investigations, it could soon be
realised, is that these did not reflect the direct
experiences of practitioners in such endeavours. Hence, the
author had to take recourse to another strategy and turned
to some of his former students employed by such
establishments. The items in the questionnaire were
carefully designed so as to elicit a positive feedback on

what they have been doing pertaining to mathematics against
the background of initial equipment, their early exposure to
mathematics, subsequent acquisition (maybe a sort of on-the-
job-education), what they would like to see in mathematics
courses, elements reflecting requirements of the day vis a
vis job-specific capabilities, and prerequisites of
mathematicians in different fields of human endeavour. The
answers to such queries appear to be, in many ways, helpful
in our bid to make mathematics a truly serviceable
proposition. Even though one does not come across a
consensus running through the answers, the overall concern
does certainly stand out as something overriding. The
perennial quest for rationality, abstraction,
generalization, etcetera as spinoff of mathematical
education continues unabated as do the banal references to
words like 'practical', 'daily-life-oriented' mathematical
material. With all such tendencies for tinkering exercises,
there appears to be now a shift in attitude and outlook for
mathematics which one could hardly have guessed a decade
ago; for example, it used to be an obsession with employers
to look for managerial finesse bereft of mathematical
content, but now they would prefer to opt for a management
scientist equipped with a fairly good level of mathematics.
Indeed, an industrial view of mathematics is fast coming up
so as to enable the category of those in charge of
innovation and development, obviously on scientific lines,
to come to the forefront; this will certainly be reinforced
if there is a continuing demand for sophistication. A
futuristic view of industry, particularly of the engineering
industry, is often being reckoned; as a matter of fact, the
development of new techniques, it is held by many, can be
facilitated by a sustained training that produces a
mathematically oriented engineer. The current pattern of
instruction in engineering-mathematics content could hardly
go unscathed. Current technology can scarcely make strides
for that of tomorrow if mathematical education is not given
a face lift - that is what many feel. There is a distinct
category of employers in industries who are worried about
fundamental changes that computer technology may bring
about, followed by an augmented volume of applications which
can hardly make any headway if they are not handled by
people with a profound understanding of mathematics.
Mathematical modelling has acquired a new character on
account of the onslaught of the developments of computer
technology which cannot dispense with mathematics; computer
literacy in industry has nearly come to be identified with
mathematical modelling. Many incumbents in industrial
enterprises wish to have mathematics courses in engineering

education permeated with a good content of practical computing and mathematical modelling.

In brief, one can say that there is a greater recognition than ever before about the societal role of mathematics. One could not anticipate a decade ago that mathematics would emerge as a productive force and would be at the service of society with a greater commitment than that which could be envisaged earlier. The words 'mathematical modelling', often seemingly industrial jargon have, of late, acquired a distinctive connotation, and have in a substantial manner extended the activity of mathematics. That mathematics can participate in activities on socio-economic programmes in serviceable ways is a realisation in contemporary times. The image of mathematics in the eyes of users is inordinately high and it is being increasingly felt, in a developing country like ours, that one can bypass mathematics at one's own disadvantage if understanding and insight into the situation is to be a studied ethos of handling societal affairs. Situational compulsions warrant in a big way the use of mathematics and that, too, in great depth. A mathematics graduate is not required by the employer to change his profession. Industry is happy to have him as a mathematics specialist but maybe not always in a professional sense. Mathematics as a service subject can assume a domineering role only if training in mathematics provides students with 'professionalism' which ought to be, doubtless, different in complexion and tenor, on account of the intrinsic character of mathematics. The service character of mathematics is evolving; its linearity is fast tapering off. Mathematics as an interlacing element has to be focussed if it is to serve a vastly diverse and steadily increasing community of users. Diffusive and dispersive characteristics ought to be reflected in any curriculum of mathematics cast in the crucible of the growing gamut of applications and of applicable mathematics. There are indications, hopeful enough, that there are going to be spill overs in structural aspects so that the pursuit of mathematics becomes an institutionalized and ongoing activity. The fallout of such thoughts in extra-mathematical circles is yet to be mirrored at the instructional level and the sooner it is done, the quicker will be the symbiosis of mathematics and mathematical sciences on a continuing basis, vide Sinha [3], [4], [5].

REFERENCES

1.   D.K. Sinha, Employment opportunities for mathematical manpower, Mathematica, 3, 1974, pp 28-40.

2. D.K. Sinha, Teaching of mathematics for users of
    mathematics, <u>Mathematics Teacher</u>, 11, 1976, pp 442-446.
3. D.K. Sinha, Teaching for users of mathematics, <u>Indian
    Journal Mathematics Teaching</u>, 4, 1977, pp 54-56
4. D.K. Sinha, Mathematization for mathematics teaching,
    <u>ACEID Newslatter</u>, (UNESCO Regional Office, Bangkok), 22,
    1981, p 12.
5. D.K. Sinha, Mathematics as a cultural entity - in
    retrospect, <u>Interdisciplinary Science Reviews</u> (in
    press).

SUR L'ABUS DES MODELES MATHEMATIQUES

J. Tonnelat
Université Paris-Sud, Orsay, France

Le texte ci-dessous est constitué de larges extraits d'un article paru dans la revue "Fundamenta Scientiae" Vol. 7 (1986) p.89-118. Les développement omis sont signalés par le symbole [...].

The text below is a partial reprinting of an article which appeared in "Fundamenta Scientiae" Vol. 7 (1986) p.89-118. Omitted parts are indicated by [...].

ABSTRACT

[On the Abuse of Mathematical Models] It is impossible to know what are "in reality" observable systems. One can only know the properties of the representations conceived for it. In a first step, these representations are models constituted of objects similar to well-known objects. These models are called physical models. Proceeding to accurate previsions entails passing to mathematical models. This passage requires that the components are assimilated to mathematical concepts. Therefore, it is not a translation into another language, but a transposition into another kind of completely abstract representation. This kind of model allows us to carry out forecasting calculations, to use more rigorous reasoning and makes possible a better analysis of the observable properties of the investigated system. But the theoreticians are wrong when they estimate it is not useful to take into account arguments founded on physical models which are not consistent with the mathematical representations familiar to them.

A mathematical model may be efficient without bringing really an explanation. Nevertheless, a mathematical model, like any representation, is a simplification which does not usually bear all the properties corresponding to those of the system, it may also sometimes possess properties from which tne system is deprived. Moreover, the correspondence between its properties and those of tne physical model may be defective. Examples are given of mathematical models offering these defects. A mathematical model may incite to injustified comparisons. Many are fruitless.

All the properties of a system, and especially all their transformations, are not expressible by means of mathematical relations. The description of tne physical model, which is an obligatory intermediary, can be made only in discursive language. A mathematical model is a rigid frame which cannot take into account tne changes in the nature of the components. Mathematical reasoning has been the determining process of the discovery of new properties only in the range of elementary particles. Mathematical reasoning is only one of the ways, as imperfect as any other, of explaining some aspects of the behaviour of the observable systems.

SOMMAIRE

- Vaine poursuite de la réalité.
- Modèle physique et modèle mathématique.
- Règne dictatorial des mathématiques.
- Défauts des modèles mathématiques :
   - Modèles efficaces mais de pouvoir explicatif discutable
   - Correspondance défectueuse entre les propriétés du modèle mathématique et celles du modèle physique
   - Limites du modèle mathématique
   - Incitation à des analogies intempestives
   - Stérélité de nombreux modèles mathématiques
   - Théories et expressions mathématiques
- Conclusion.

Quel que soit leur domaine de recherche, il paraît indispensable à de nombreux auteurs de s'efforcer d'exposer les résultats de leur travail sous forme d'expressions mathématiques. La confiance dans l'aptitude du langage mathématique à rendre compte des phénomènes et la croyance en la nécessité d'en faire le moyen d'expression essentiel sont très généralement partagées. L'idée a même été plusieurs fois émise que le monde était constitué d'une façon conforme aux algorithmes mathématiques. Galilée disait déjà : "On ne peut le comprendre (ce livre immense de l'univers) si l'on n'apprend pas d'abord à connaître la langue dans laquelle il est écrit" ; Jeans abondait dans ce sens : "Le Grand Architecte semble être mathématicien", et Poincaré renchérissait : "La nature est la réalisation de ce qu'on peut imaginer de plus simple mathématiquement". La place prééminente attribuée aux mathématiques par Auguste Comte dans sa classification des sciences a

fortement contribué à faire croire qu'une science ne pouvait parvenir
à son plein développement que lorsque tous les phénomènes qu'elle
analysait étaient traduits en langage mathématique. Il ne faut donc pas
s'étonner de la tendance systématique à tout ramener à des relations
mathématiques, même en économie et en sociologie.

Or, il n'est pas évident que cet engouement pour la réduction
systématique des phénomènes observables à des relations mathématiques
soit bien justifiée. Il y a lieu de se demander tout d'abord ce que l'on
attend d'une description mathématique des phénomènes. Pour répondre
à cette question, il faut examiner les résultats que les hommes peuvent
espérer obtenir par une étude méthodique des phénomènes qu'ils ont la
possibilité d'observer. Ce sujet a déjà été souvent abordé. Il paraît
néanmoins nécessaire d'en reprendre l'analyse de façon à pouvoir
mettre en évidence le rapport existant entre les observations et le
modèle mathématique. Il semble que, en général, ce rapport ne soit pas
convenablement perçu par les auteurs.

LA VAINE POURSUITE DE LA REALITE

Les hommes essayent d'échapper aux désagréments qui proviennent de
leur environnement et à profiter au mieux des avantages qu'il leur est
possible d'en tirer. Pour parvenir à ce but, il faut qu'ils connaissent
les modalités du déroulement des phénomènes qu'ils observent, ou, comme
ont dit habituellement, qu'ils les "comprennent". Il est difficile de
préciser le sens qu'il faut donner à ce mot.Cependant, on constate que,
pratiquement, comprendre un phénomène, c'est d'être capable de l'"expliquer", et l'analyse des "explications" données dans la littérature
montre qu'expliquer consiste à traduire le déroulement d'un phénomène
compliqué en termes applicables à des phénomènes familiers perceptibles
à nos sens. C'est donc, en fait, assimiler un phénomène apparemment
mystérieux, en ce sens que ses modalités ne sont pas directement
perceptibles, à un phénomène ou à un ensemble de phénomènes qui
paraissent simples et "normaux".

L'opération n'est pas directement faisable lorsqu'on a affaire à
des phénomènes très complexes. Dans ce cas, on procède à une assimilation intermédiaire à un phénomène moins compliqué déjà expliqué
antérieurement. Un tel phénomène plus simple peut servir à l'explication
de plusieurs phénomènes compliqués différents, de sorte que, pratiquement, expliquer un phénomène consiste le plus souvent à le ramener à un
phénomène plus général, le qualificatif de général indiquant que ce
phénomène sert de référence pour l'explication de plusieurs types de
phénomènes différents. De proche en proche, on ramène l'explication d'un
phénomène quelconque à celle d'un phénomène familier. On dit alors que
le phénomène est "compris". A partir du moment où des phénomènes sont
compris, on peut souvent prévoir les modalités de leur évolution quand
ils s'amorcent spontanément, parfois alors les empêcher de se dérouler
ou de modifier les cours, et, surtout, pour un grand nombre d'entre eux,
les provoquer dans des conditions favorables aux desseins des hommes.

On pense courramment qu'une bonne explication permet de savoir en
quoi consiste un phénomène "en réalité". Or, l'idée que l'on puisse

arriver à connaître le monde "réel" est illusoire. Tout homme constate
qu'il se trouve environné d'objets, vivants ou non, qui le heurtent ou
le caressent, dont la rencontre est bénéfique, si par exemple , ils lui
servent de nourriture ou augmentent son confort, ou est au contraire
déplorable s'ils le gênent ou le blessent. Il est donc difficile de ne
pas admettre qu'il existe autour de chacun un monde que l'on qualifie de
"réel" parce que ses effets sur nous ne paraissent pas être le fruit de
notre imagination. Nous connaissons ce monde par l'effet de ses propriétés sur nos sens. On doit se demander dans quelle mesure l'interprétation de nos observations sur ces systèmes nous renseigne sur leurs
propriétés "réelles".

      La description de nos observations ne peut se faire qu'en termes
déjà connus, donc en assimilant les propriétés des objets observés à
celles des objets courants. Pratiquement, une telle description est
l'exposé du comportement qu'auraient des objets *analogues* à des objets
familiers qui seraient placés dans des situations *ressemblant* à des
situations familières. Il peut cependant s'avérer nécessaire d'introduire
de nouveaux termes, ou de donner de nouveaux sens à des mots usuels,
afin de caractériser des propriétés qui diffèrent notablement de celles
des objets familiers. Le besoin d'introduire ces innovations est
révélateur du fait que cette assimilation de propriétés de systèmes
complexes aux propriétés des systèmes usuels n'est pas vraiment satis-
faisante, que les observateurs en aient eu conscience ou non. A la
réflexion, il paraît en effet évident, par exemple, que les propriétés
entraînant l'orientation d'une pierre d'aimant dans le champ terrestre
ou l'attaque de la craie par le vinaigre ne sont pas de même nature que
celles d'objets usuels comme une girouette ou de petites boules dont les
modalités d'association sont susceptibles d'être modifiées. L'explication
de phénomènes complexes observables par des analogies avec des phénomènes
familiers laisse manifestement à désirer. Elle ne rend pas vraiment
compte de leur nature, donc ne permet pas de savoir ce qu'ils sont "en
réalité". En fait, les explications consistent à élaborer une
représentation par des objets familiers, ou définis par leurs analogies
avec des objets familiers, de systèmes dont on ne connait que les
manifestations de certaines de leurs propriétés, ces manifestations
étant celles que font apparaître les moyens d'observation. Il est
absolument impossible de savoir dans quelle mesure les propriétés de
ces représentations sont effectivement assimilables à celles des
systèmes observées, en dépit des soins que l'on a pu prendre pour
affiner cette représentation. *En aucun cas on ne peut connaître les
propriétés d'un système "réel", on ne peut arriver à connaître que celles
de la représentation que l'on s'en fait.*

      Quel que soit le problème physique considéré, tout observateur est
inévitablement contraint de décrire uniquement le comportement de
systèmes imaginaires, qualifiés de "modèles", dont le comportement est
censé être *analogue* à celui des systèmes étudiés, dans la mesure où
ceux-ci sont observables. Il n'est pas possible de décrire un monde
"réel", d'expliquer ce qu'il est "en réalité" parce que cette expression
n'a pas de sens. Ce qui est étudié n'est jamais un système réel, mais
toujours un modèle construit à partir de l'interprétation d'observations.

Le modèle n'a en commun avec la "réalité" que de donner des résultats considérés comme *équivalents* dans le cadre des observations effectuées. Un observateur peut seulement dire que, pour lui, *tout se passe comme si* le système étudié jouissait des propriétés attribuées au modèle, mais les propriétés propres du système restent inconnues.

MODELE PHYSIQUE ET MODELE MATHEMATIQUE

Donc, tout le monde étudie des modèles et jamais personne ne peut affirmer qu'un modèle est parfaitement représentatif du système étudié.

La compréhension d'un phénomène consiste dans un premier temps à s'en faire une représentation résultant d'une assimilation de ses constituants à des objets directement perceptibles à nos sens, tels que des boules ou des tiges, animés de mouvements mécaniques et interagissant éventuellement à distance. Les modèles "construits" à partir de ce genre d'assimilation se présentent à l'esprit sous un aspect analogue à celui des systèmes qui sont le siège des phénomènes que nous pouvons observer directement avec nos sens. Je qualifierai ces modèles de physiques. Dans les cas favorables, il est possible de construire des maquettes qui "matérialisent" ces modèles. A tout le moins, un modèle de ce type apparaît à l'esprit comme susceptible d'être construit.

Les capacités de l'esprit humain sont insuffisantes pour appréhender toutes les propriétés d'un système. En conséquence, un modèle est toujours une représentation simplifiée qui néglige systématiquement les propriétés qui *paraissent* secondaires dans l'étude du phénomène observé. L'adoption, inévitable, d'un modèle représente l'avantage de substituer au système observé un système dont ont été élaguées des propriétés dont la prise en considération compliquerait inutilement une explication que l'on souhaite aussi simple que possible. La conception d'un modèle implique donc une schématisation. Si la simplification correspondante est judicieuse, l'examen du modèle rend possible la mise en évidence et l'analyse de propriétés qui n'avaient pas été décelées directement par la simple observation du système parce que celui-ci se présentait comme un ensemble trop complexe.

Cependant, une représentation physique ne permet généralement pas d'expliquer complètement le comportement d'un système, ni surtout de préciser les déplacements ultérieurs de ses constituants. Une telle étude demande des calculs mathématiques.

Le recours au raisonnement mathématique ne peut s'effectuer que si les propriétés du système, ou plus exactement, celles de la représentation que l'on s'en fait, donc celles du modèle physique, ont été préalablement schématisées d'une façon particulière consistant, par exemple, à assimiler certains constituants à des points possédant une masse ou une charge, entre lesquels s'exercent des forces définies. Les représentants de tous les éléments du système sont alors des êtres mathématiques. L'expression mathématique des influences réciproques s'exerçant entre ces êtres figuratifs constitue une représentation de celles des propriétés physiques du système étudié qui peuvent être exprimées par des relations mathématiques. Cette représentation est un modèle mathématique. Il faut bien remarquer que c'est une représentation

complètement abstraite. Le recours aux mathématiques n'est donc pas une
simple traduction dans une langue mieux adaptée, c'est une *transposition*
dans un mode de représentation qui jouit de propriétés d'une nature
complètement différente. Cette manière de faire est le plus souvent
justifiée par la possibilité d'établir une correspondance entre les
propriétés des êtres mathématiques ainsi introduits et celles des
objets figuratifs du modèle physique.

L'obtention d'un modèle mathématique est généralement considérée
comme l'aboutissement idéal de la recherche d'une représentation. La
raison primordiale en est qu'un tel modèle autorise un traitement
quantitatif de son comportement qui permet de prévoir son évolution
avec précision et de confronter aux valeurs expérimentales les valeurs
des grandeurs calculées à partir de la "théorie". La théorie dans ce
sens est l'exposé des propriétés du modèle mathématique et aussi, et
même très souvent surtout, de celles du modèle physique sur lesquelles
le premier est fondé. Une discordance marquée indique sans doute
possible que les explications doivent être reprises sur une autre base.

Une deuxième raison est que ce modèle n'a que des propriétés
parfaitement définies. Il permet des raisonnements rigoureux. Dans les
algorithmes mathématiques, le principe d'identité et de contradiction
est parfaitement valable : A est A, A n'est pas non-A, il n'y a pas
de milieu entre A et non-A.

Enfin, et cet avantage est d'une importance considérable, on
constate que les algorithmes applicables aux modèles mathématiques
permettent de mettre en évidence des propriétés qui n'étaient pas
apparentes avec beaucoup plus d'efficacité que les modèles physiques,
grâce à la puissance d'analyse des raisonnements mathématiques.

L'exploitation d'un modèle mathématique conduit très souvent à
constater que le modèle physique à partir duquel il a été établi laisse
à désirer et qu'il est nécessaire de le modifier pour l'étude de
certains phénomènes, même si le premier rend mieux compte d'autres
phénomènes. Le nouveau modèle physique servira de base à de nouveaux
développements mathématiques qui pourront éventuellement conduire à
leur tour à une nouvelle révision et ainsi de suite. La recherche de
l'interprétation des résultats numériques peut aussi conduire, non à
apporter des corrections au modèle physique initial, mais à lui en
substituer un autre notablement différent. C'est ainsi que, pour l'étude
à petite échelle de nombreux phénomènes, on a été conduit à la conclusion qu'un solide apparemment continu à grande échelle devait être
considéré comme un milieu presqu'entièrement vide dans lequel évoluent
de minuscules corpuscules situés à des distances relativement consi-
dérables les uns des autres. Ce qui n'apparaît pas dans cette image est
que ce vide est le siège d'interactions intenses. Cet exemple montre
bien que, si ce qu'on appelle dans un tel cas la *réalité* est une image
du système rendant convenablement compte de certaines propriétés, ici
une répartition des constituants dans l'espace, cette image est
fallacieuse, parce qu'elle ne prend pas en compte une propriété
essentielle du système qui "explique" et cette répartition et le
comportement du solide confronté à des phénomènes perturbateurs.

Quoi qu'il en soit, dans tous les cas, un modèle mathématique

correspond à un modèle physique, puisqu'un lien entre expressions
mathématiques et résultats expérimentaux ne peut exister que si une
correspondance est établie entre des grandeurs abstraites et des
propriétés d'entités que l'esprit peut se représenter. Même si ce
dernier est retouché en fonction des améliorations inspirées par le
modèle mathématique, les deux types de modèles restent complémentaires.

Pour mémoire, il y a lieu de rappeler un inconvénient grave que
présentent tous les modèles mathématiques. C'est celui d'exiger un
langage ésotérique. Cependant, c'est justement la transposition dans
ce langage qui permet de recourir à un mode de raisonnement plus
puissant. Il serait donc évidemment souhaitable que tout expérimenta-
teur soit capable de le comprendre. Devant l'impossibilité de réaliser
ce rêve, les théoriciens devraient toujours prendre soin d'exprimer
en langage clair leurs hypothèses de départ, celles qu'ils sont
amenés à faire en cours de route et leurs conclusions. Ce n'est
malheureusement pas souvent le cas.

REGNE DICTATORIAL DES MATHEMATIQUES

A l'heure actuelle, une théorie physique ne s'exprime jamais en
termes purement discursifs. L'exposé complet des théories implique
toujours un appareil mathématique, généralement important, qui a donné
la possibilité d'approfondir les propriétés du système et qui seul
permet des vérifications quantitatives. On comprend donc que
d'Espagnat [1] ait pu dire que "l'outil mathématique... s'est progressivement révélé le plus efficace -et de beaucoup- pour la synthèse dont
il s'agit (la systématisation de l'ensemble des connaissances concernant le monde réel)". Il ajoute même plus loin, en considérant cependant un domaine particulier : "Même quand les descriptions des particules
élémentaires font image, ce qui leur confère leur valeur, et on doit
même dire leur substance, ce n'est pas l'imagerie spatiale qu'elles
évoquent (cette dernière est toujours en partie trompeuse), c'est la
mathématique qui les soutient". Il est de fait que les connaissances
acquises sur la représentation que l'on se fait de la structure
élémentaire de la matière n'ont pu être obtenues qu'au prix de raisonnements mathématiques très élaborés. Ce succès n'a pas peu contribué
à inciter à croire à la nécessité de proposer, et même d'imposer, des
modèles mathématiques dans tous les domaines.

Beaucoup de théoriciens ne perçoivent une propriété d'un système
que lorsque celle-ci s'énonce sous forme mathématique. Ils ne sont
jamais convaincus par un raisonnement discursif. Peut-être faut-il
attribuer leur répulsion pour ce mode d'expression, d'une part au fait
qu'un discours ne peut prétendre à une aussi grande concision qu'un
raisonnement mathématique, et, d'autre part, à ce qu'il est plus
difficile de juger de sa rigueur. Quelle qu'en soit la raison, cette
répulsion est telle que les spécialistes des développements mathématiques ne jugent pas utile de prendre en considération des critiques
fondées sur des modèles physiques qui ne sont pas conciliables avec les
représentations mathématiques qui leur sont familières. Ils se croient
alors autorisés à censurer les analyses en langage ordinaire. Il

règne ainsi un véritable terrorisme de fait, déjà dénoncé [2,3] ; mais qui ne le sera jamais trop, qui interdit d'avoir l'audace d'avancer une interprétation digne d'être qualifiée de scientifique si elle n'est pas essentiellement mathématique. Chez la plupart des théoriciens, ce terrorisme est inconscient puisque leur forme d'esprit ne leur permet d'apprécier que certains types de raisonnements abstraits, et que, en conséquence, il leur paraît évident que ce type de raisonnements s'impose dans tous les cas.

Les théoriciens semblent obnubilés par les expressions mathématiques. L'exposition d'un phénomène est devenu pour eux essentiellement l'énoncé des relations mathématiques qui en rendent compte (voir, par exemple, les exposés récents des principes de la RMN). Le modèle physique est devenu presque superfétatoire, alors qu'il est l'intermédiaire obligé entre le modèle mathématique et les observations.

L'importance primordiale que les théoriciens accordent au modèle mathématique se traduit souvent de façon curieuse dans les modes d'exposition qu'ils utilisent. Ils s'expriment en effet comme s'ils attribuaient un effet causal à la construction de l'esprit élaborée à partir d'études expérimentales. Ils en viennent alors à présenter les phénomènes observables comme des conséquences de l'existence d'un modèle qui est une pure création de l'esprit. C'est ainsi que l'on rencontre des expressions comme : "Le hasard et l'irréversibilité sont tous deux des conséquences de la structure des équations du mouvement" [4], ou "ce comportement (de l'hélium liquide sortant spontanément d'un récipient au voisinage du zéro absolu) est une conséquence de la mécanique quantique" [5]. Il est bien évident que l'on ne doit pas prendre au pied de la lettre cette manière d'exposer les choses, et qu'il faut admettre que c'est parce qu'elle facilitait leur exposé que les auteurs l'on adoptée. Mais on ne peut s'empêcher de constater qu'elle est révélatrice d'un état d'esprit attribuant une primauté absolue à la représentation mathématique.

Or, s'il est incontestable que la représentation mathématique apporte des possibilités d'analyse irremplaçables, elle peut, elle aussi, être certainement trompeuse dans tous les domaines de la physique, même dans celui des particules élémentaires. Les modèles mathématiques peuvent en effet présenter un certain nombre de défauts graves. Il en résulte que le recours à ce type de modèle n'est pas toujours fait à bon escient. Il peut arriver aussi qu'un tel modèle ne soit pas convenablement exploité. Avant d'en donner des exemples, il y a lieu d'attirer l'attention sur le fait qu'un modèle mathématique peut jouer un rôle efficace alors même qu'il n'apporte pas vraiment d'explication au sens défini plus haut.

RESERVES A FAIRE SUR CERTAINS MODELES MATHEMATIQUES EFFICACES MAIS DE POUVOIR EXPLICATIF DISCUTABLE

Pour qu'il puisse avoir un pouvoir explicatif, un modèle physique doit jouir de propriétés analogues à celles d'objets familiers. Or, ce caractère peut laisser à désirer. C'est ainsi que l'on n'est pas

parvenu à trouver de modèles physiques satisfaisants pour toute une
catégorie très importante de phénomènes, ceux qui concernent les
interactions à distance sans intermédiaire apparent. [...]

CORRESPONDANCE DEFECTUEUSE ENTRE LES PROPRIETES DU MODELE MATHEMATIQUE ET CELLES DU MODELE PHYSIQUE

Quel que soit le modèle adopté, pour qu'il soit acceptable, il
faut qu'il jouisse de toutes les propriétés correspondant à celles que
l'on désire étudier dans le système observé. C'est là la qualité
essentielle d'un modèle (Thom [8]). Or, bien que tous les auteurs
soient, au moins implicitement, d'accord sur ce point, il n'est pas
rare que des modèles ne satisfassent pas pleinement à cette condition.
D'autre part, il peut arriver qu'un modèle présente le défaut contraire,
c'est-à-dire qu'il jouisse de propriétés que ne possède pas le
système qu'il est censé représenter ; l'éventualité de l'existence
de ce défaut est presque toujours ignorée. Enfin, il faut que la
correspondance entre les propriétés du modèle et celles du système
observé soit correctement établie. Une correspondance incorrecte se
confond souvent avec l'un des deux défauts indiqués. Nous allons voir
une série d'exemples de modèles présentant ces différents défauts. [...]

1 - Modèles mathématiques jouissant de propriétés dont le modèle
   physique est dépourvu. [...]

2 - Ajustement incorrect des propriétés du modèle mathématique à celles
   du modèle physique. [...]

3 - Modèle mathématique ne rendant pas correctement compte des propriétés
   du système.

La puissance et la rigueur du raisonnement mathématique incitent
à introduire un modèle mathématique le plus rapidement possible lors
de l'analyse du fonctionnement d'un système. Mais il n'est pas
exceptionnel que la hâte mise pour passer à la mathématisation conduise
à des systématisations injustifiées. [...]

LIMITES DU MODELE MATHEMATIQUE

L'affirmation de Claude Bernard suivant laquelle "Il n'y a de
scientifique que du mesurable" a beaucoup contribué à faire croire que
tout devait se ramener à des expressions mathématiques. Pourtant, il
est évident que nombre d'explications scientifiques ne peuvent se
traduire par des relations mathématiques entre grandeurs mesurables. De
telles relations concernent nécessairement des grandeurs bien définies.
Or, d'une part, les propriétés essentielles des objets observables sont
loin de pouvoir être exprimées au moyen d'expressions mathématiques, et,
d'autre part, ainsi que nous l'avons déjà noté à propos de la thermodynamique statistique, les caractères qui servent à les définir sont
sujets à des changements de nature ; ces objets se transforment,
fusionnent ou se scindent ; les interactions entre eux sont modifiées :

dans certains cas simples, des raisonnements mathématiques peuvent tenir
compte de ces transformations (théorie des bifurcations, théorie des
catastrophes). Le plus souvent, la nature des objets est si complexe
et les transformations qu'ils subissent sont tellement nombreuses, et
surtout tellement compliquées, que l'on ne peut songer à les décrire
autrement qu'en utilisant le langage ordinaire.[...]

INCITATION A DES ANALOGIES INTEMPESTIVES

Il arrive très fréquemment que des modèles physiques représentant
des phénomènes de natures complètement différentes conduisent à
l'établissement d'expressions mathématiques identiques.

En général, tout le monde est bien persuadé que l'identité de
la forme mathématique représentative n'implique l'existence d'aucun
lien nécessaire entre les grandeurs concernées dans les deux
phénomènes se traduisant par la même équation.[...]

Or, à la suite de Brillouin [15], la plupart des auteurs croient
fermement qu'il existe un lien étroit entre l'entropie et
l'"information" [...], il n'y a aucune raison pour qu'il existe un lien
entre entropie thermodynamique et information.[...]

Il 'est pas rare que l'on rencontre un autre genre de rapproche-
ment injustifié. Ces rapprochements sont ceux qui sont établis entre
les résultats obtenus par le traitement mathématique de problèmes
simples et certains caractères observables de phénomènes compliqués,
et particulièrement de phénomènes biologiques.[...]

STERILITE DE NOMBREUX MODELES MATHEMATIQUES

On peut se demander si ce n'est pas avant tout le prestige des
mathématiques bien plus que l'espoir de pouvoir utiliser la puissance
d'analyse des algorithmes ainsi rendus accessibles qui incite beaucoup
d'auteurs à développer au maximum toutes les conséquences théoriques
qui peuvent être déduites d'un modèle mathématique. En effet, il
arrive fréquemment que les auteurs ne paraissent pas se soucier des
possibilités d'en déduire des conséquences pratiques.[...]

Or une mathématisation n'est vraiment intéressante que si elle
permet effectivement :
- soit une expression plus claire grâce à l'utilisation d'une notation
  plus condensée ;
- soit une analyse plus approfondie et la mise en évidence de
  propriétés intéressantes ;
- soit, à tout le moins, des calculs numériques dont les résultats,
  confrontés aux résultats expérimentaux, seront des tests de la
  validité de la représentation adoptée.

THEORIE ET EXPRESSIONS MATHEMATIQUES[...]

CONCLUSION

Il n'est pas question de nier l'intérêt spéculatif des développements abstraits, qui sont d'ailleurs parfois susceptibles de retombées
pratiques. Cependant, on doit constater que, en ce qui concerne les
phénomènes physiques et, *a fortiori*, les phénomènes biologiques qui
n'en sont que des manifestations particulières, la contribution des
développements purement mathématiques à la progression de nos connaissances n'a été véritablement fondamentale que dans le domaine des
particules élémentaires. Les phénomènes correspondants sont des phénomènes "simples" concernant des systèmes soigneusement sélectionnés
dans des conditions bien définies. Mais, pour l'analyse des phénomènes
naturels, dont les paramètres expérimentaux ne sont pas maîtrisables,
les développements mathématiques ne peuvent jouer un rôle que pour des
étapes limitées bien qu'essentielles. Il est donc bien justifié de
déployer de grands efforts pour décrire une partie aussi grande que
possible des phénomènes par des relations mathématiques. Mais on se
heurte dans leur application à des barrières en raison même de la
structure de la représentation mathématique : certaines des propriétés
des systèmes observés ne peuvent absolument pas être traduites par des
relations mathématiques et le modèle mathématique peut être incapable
de s'adapter à des modifications du système. Les auteurs semblent
s'efforcer de ne pas remarquer cette limitation des capacités d'un
modèle mathématique. Elle est pourtant capitale.

D'autre part, la représentation mathématique est susceptible de
faillir à son rôle, soit que le modèle mathématique présente des
propriétés supplémentaires par rapport au système observé, soit que des
propriétés essentielles lui fassent défaut, soit encore que la
correspondance établie entre ses propriétés et celles du système soit
défectueuse.

Il n'est pas exceptionnel que d'élégants développements fassent
penser à des contes de fées. Ils exposent des résultats séduisants
auxquels il est agréable de croire, mais ces résultats ne seraient
corrects que si l'on admettait une opération irrationnelle. Ils
impliquent en effet que, par un processus différent mais aussi extraor-
dinaire que la transformation d'une citrouille en carrosse, un modèle
mathématique impropre peut, d'un coup de baguette et sans qu'il soit
changé, être rendu apte à représenter des faits qu'il ne traduit pas.

Par ailleurs, la confiance dans les vertus de tout modèle mathématique incite à des rapprochements sans fondement ou à des développements sans intérêt.

L'abus des mathématiques est particulièrement criant dans leurs
applications hardies à l'économie, au point que des spécialistes
reconnaissent en privé que des théories élaborées, qui inspirent
d'autant plus de respect au profane qu'elles sont plus ésotériques, sont
en fait des colosses aux pieds d'argile. On voudrait être sûr qu'il
n'en va pas parfois de même de certaines théories en physique et surtout
en biologie. Les auteurs passent toujours beaucoup plus de temps à
analyser les propriétés mathématiques de la représentation adoptée
qu'à en discuter les fondements. Ils se lancent sans précautions dans

des développements aux bases discutables avec l'assurance des grands
prêtres s'appuyant sur le dogme. Ils trouvent même étrange que l'on
puisse mettre en doute des idées reçues et ne veulent pas entendre
quand on dit que le roi est nu.

La représentation mathématique est seulement l'un des moyens
utilisables pour parvenir à la "compréhension" du monde extérieur. Il
est particulièrement puissant, mais il peut se montrer inefficace, et
même nocif s'il est utilisé mal à propos. Pour des raisons d'"économie
de pensée", il serait évidemment très agréable de pouvoir exprimer
d'une façon aussi concise que possible les principes fondamentaux
communs à toutes les sciences étudiant les phénomènes naturels (de la
physique à la biologie, en passant par l'astronomie). Il n'est
peut-être pas illusoire d'espérer atteindre ce but. Mais cela ne sera
possible que si l'on utilise des moyens adéquats. S'imaginer *a priori*
que tous ces principes pourront s'exprimer exclusivement, ou même
essentiellement, dans le langage mathématique revient à s'interdire de
chercher une représentation complète du cadre dans lequel les représentations mathématiques sont valables. En effet, la description de
ce cadre et des contraintes qu'il implique, c'est-à-dire en fait la
description du modèle physique, ne peut se faire qu'en langage
discursif et même souvent en faisant appel à des dessins. Il est vrai
que l'explication d'un phénomène n'est pas satisfaisante quand ce
cadre reste vide, mais ce n'est pas une raison pour le bourrer à toute
force de relations mathématiques. Il n'est pas exceptionnel que le
commentaire d'un croquis en langage ordinaire soit plus explicite
qu'une équation.

Il paraît nécessaire de rappeler une fois de plus que, pour
l'analyse des phénomènes observables, le raisonnement mathématique est
un bon serviteur, qui procure les outils pouvant éventuellement
conduire à des explications satisfaisantes, mais un mauvais maître,
qui devient trop souvent un tyran imposant un cadre de pensée nuisible
à la réflexion.

REFERENCES

[1] d'Espagnat B., A la recherche du réel, Gauthiers-Villars, Paris
     (1979)
[2] Delattre P., Systèmes, Structures, Fonction, Evolution, Maloine-
     Doin, (1971)
[3] Nordon D., Vous avez dit mégalomane ? Fund. Sci. 2, 67 (1981)
[4] Prigogine I., Physique, temps et devenir, Masson, Paris (1980)
[5] Balibar S., La Recherche 11, 712 (1980)
[8] Thom R., Modélisation et scientificité, in *Elaboration et justifica-
     tion des modèles* (Edités par Delattre P. et Thellier M.) Maloine
     (1979).
[15] Brillouin L., *La science et la théorie de l'information*, Masson,
     Paris (1959).

FINAL STATEMENT

Mathematics is of increasing importance in all sciences and in everyday life. It is an essential part of the general culture needed by every citizen in order to understand our world and treat information and data with a critical mind. It is already an essential tool for many professions and will become necessary for many more in the future.

Mathematics has, therefore, to be taught to many students whom mathematicians have not considered before - to students of subjects as widely differentiated as home economics and biology. Even in the fields where a mathematical education is a tradition - such as physics and engineering - many changes are necessary. Advances in mathematical and computational tools make mechanical techniques and even skills less important than before. Mathematical understanding becomes even more crucial when students and professionals use computers, symbolic manipulation systems, computer graphics and other kinds of new technology. For the same reasons continuing education demands an increasingly important role. The successful design of mathematical courses to meet these needs requires an increased degree of understanding and cooperation between mathematics teachers and those in other disciplines.

All mathematicians must be aware that the future of mathematics as a science depends on the way they respond to these new needs coming from other disciplines and from society as a whole.

Public opinion and governments should be made aware of the urgency of meeting these new needs. The status of service teaching and service teachers must be improved. New appointments, new means and increased resources are vital.